COOPERATIVE-LEAR...

HOLT, RINEHART AND WINSTON

A Harcourt Classroom Education Company

Austin • New York • Orlando • Atlanta • San Francisco • Boston • Dallas • Toronto • London

To the Teacher

Algebra 2 Cooperative-Learning Activities contain one-page blackline masters for each of the 93 lessons in *Algebra 2*. These masters provide structured activities for students to perform in small groups or in pairs to reinforce the mathematical content in the lessons. Directions designate specific roles or responsibilities that facilitate student cooperation and participation.

Copyright © by Holt, Rinehart and Winston

All rights reserved. No part of this publication may be reproduced or transmitted in any form or by any means, electronic or mechanical, including photocopy, recording, or any information storage and retrieval system, without permission in writing from the publisher.

Teachers using ALGEBRA 2 may photocopy complete pages in sufficient quantities for classroom use only and not for resale.

Photo Credit
Front Cover: Tom Paiva/FPG International.

Printed in the United States of America

ISBN 0-03-054078-X

 2 3 4 5 6 7 066 02 01 00

Table of Contents

Chapter 1	Data and Linear Representations	1
Chapter 2	Numbers and Functions	9
Chapter 3	Systems of Linear Equations & Inequalities	16
Chapter 4	Matrices	22
Chapter 5	Quadratic Functions	27
Chapter 6	Exponential and Logarithmic Functions	35
Chapter 7	Polynomial Functions	42
Chapter 8	Rational Functions & Radical Functions	47
Chapter 9	Conic Sections	55
Chapter 10	Discrete Mathematics: Counting Principles and Probability	61
Chapter 11	Discrete Mathematics: Series and Patterns	68
Chapter 12	Discrete Mathematics: Statistics	76
Chapter 13	Trigonometric Functions	82
Chapter 14	Further Topics in Trigonometry	88
Answers		94

NAME _____ CLASS _____ DATE _____

Cooperative-Learning Activity
1.1 Exploring Fixed and Variable Costs in Business

Group members: 3

Materials: grid paper

Roles: **Problem Maker** chooses a fixed cost and a variable cost
Table Maker represents the relationship between units and total cost in a table
Graph Maker graphs the relationship between units and total cost

Preparation: Your group will explore how a linear equation can be helpful in a business situation. You will represent cost data in tabular, graphical, and algebraic forms.

Procedures:
1. The Problem Maker decides what amount of money is needed to keep a small crafts shop open for one week. This amount is referred to as the *fixed cost*. He or she also decides how much it will cost to make one small ceramic vase. This amount is referred to as the *variable cost*.

2. The Table Maker uses this information to find the total cost, c, of making one ceramic vase. The total cost is the sum of the fixed cost and the variable cost. The Table Maker completes row 1 of the table below.

3. The Graph Maker sets up a coordinate grid and plots the ordered pair (n, c) from row 1 of the table.

4. The Table Maker and the Graph Maker complete the table and graph for $n = 2, 3, 4, \ldots, 7$.

5. As a group, write an equation that relates n and c.

   _____

6. The Problem Maker uses the equation to find the total cost of making 105 vases, 120 vases, 130 vases, and 142 vases.

   _____

Number of vases, n	Fixed cost	Variable cost	Total cost, c
1			
2			
3			
4			
5			
6			
7			

Algebra 2   **Cooperative-Learning Activity   1**

NAME _____ CLASS _____ DATE _____

Cooperative-Learning Activity
1.2 The Slope Race

Group members: 2

Materials: coin, number cube

Responsibilities: Players take turns plotting triangles determined by slopes.

Preparation: The first roll of the number cube gives a, the numerator of a slope. The second roll gives b, the denominator of the slope. Heads on the coin toss indicates a positive slope, and tails indicates a negative slope.

Procedures:
1. Choose who will go first.

2. a. Player 1 rolls the number cube. The number showing is a. Player 1 rolls the number cube again. The number showing is b. The coin toss indicates whether the player moves up a units or down a units.

   b. Player 1 puts his or her pencil at the origin. The player moves b units to the right and then a units up or down. The player marks the point so located with a small picture of his or her choice. An example might be a thimble.

3. Player 2 completes Procedure 2 just as Player 1 did. Player 2 marks the point with a picture of his or her choice. An example might be a racing car.

4. Players alternate completing Procedure 2 starting from the point marked on his or her preceding turn. If a player reaches the upper or lower edge of the grid during a turn, that player puts his or her marker at the edge of the grid.

5. The player who crosses the finish line first is the winner.

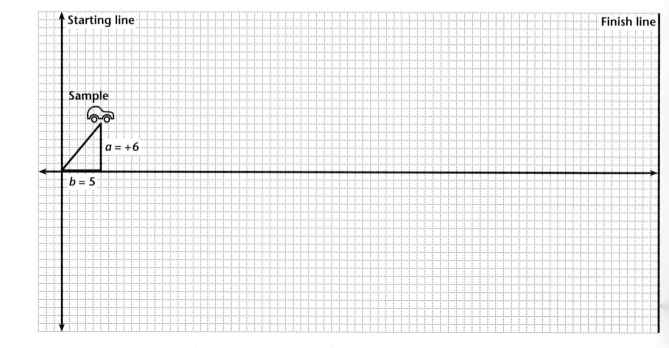

2   Cooperative-Learning Activity   Algebra 2

NAME _____ CLASS _____ DATE _____

Cooperative-Learning Activity
1.3 Counting Whole Number Solutions

Group members: 3

Materials: grid paper, two number cubes

Roles: **Equation Maker** rolls the number cubes to find a in the equation $x + y = a$

Equation Grapher graphs the equation found by the Equation Maker

Equation Solver graphs and counts the number of ordered pairs whose coordinates are whole numbers that satisfy $x + y = a$

Preparation: Your group will use experimentation and reasoning to make a conjecture about solutions to equations in two variables.

Procedures:
1. Each member chooses a role.

2. The Equation Maker rolls the two number cubes. The sum of the numbers showing is the value of a in the equation $x + y = a$.

3. The Equation Grapher graphs $x + y = a$ using a as found in Procedure 1.

4. The Equation Solver graphs each ordered pair (x, y) that is a whole-number solution to $x + y = a$. Then the Equation Solver records a and the number of solutions found in Row 1 of the table at right.

a	Number of solutions

5. Switch roles so that no member has the same role as before. Repeat Procedures 2–4. If the value of a is a value already found, the Equation Maker rolls the number cubes again until a new value of a is found.

6. Switch roles again so that everyone has a role not yet taken, and repeat Procedures 2–4. If the value of a is a value already found, the Equation Maker rolls the number cubes again until a new value of a is found. The process is continued until each member has assumed each role twice and the table is complete.

7. Work together to write a relationship between a and the number of ordered pairs (x, y) that satisfy $x + y = a$ and have whole number coordinates.

   _____

8. Work together to write a justification for the conclusion that you drew in Procedure 7.

   _____

   _____

Algebra 2   **Cooperative-Learning Activity**

NAME _____ CLASS _____ DATE _____

Cooperative-Learning Activity
1.4 Circle Measurements and Variation

Group members: 3

Materials: compass, ruler, string, rice

Responsibilities: Use the materials listed above to make discoveries about circles, circle measurement, and variation.

Preparation: In this activity, your group will draw conclusions about relationships between the radius of a circle, its circumference, and its area.

Procedures:

1. On separate sheets of paper, each member of the group draws a circle. The three circles will have radii of 1 inch, 2 inches, and 4 inches. Draw lines that divide your circle as shown at right.

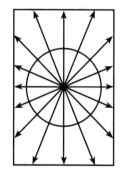

2. Wrap a piece of string around your circle exactly once. Then lay the string out along a ruler to find its length. The results should be recorded in the table.

Radius	Careful estimate of the circumference
1 inch	
2 inches	
4 inches	

3. As a group, find a direct-variation relationship between the radius of each circle and the length of the string needed to wrap around it. Write the relationship that you discover below. What is the constant of variation?

_____

4. The member who drew the 1-inch circle covers the circle completely with rice. The member who drew the 2-inch circle moves the rice to cover one or more of the one-sixteenth sections of that circle. The member who drew the 4-inch circle moves the same amount of rice to one or more of the sections of that circle. Complete the table below.

Radius	Fraction of circle covered by rice
1 inch	
2 inches	
4 inches	

5. As a group, compare the area of a circle with a radius of r to the area of a circle with a radius of 1.

_____

NAME _____ CLASS _____ DATE _____

Cooperative-Learning Activity
1.5 Age and Leisure Activities

Group members: 4

Materials: grid paper, graphics calculator

Roles: **Plotter** uses graphics calculator to sketch a scatter plot of the data

Coordinate Maker makes coordinate pairs from data taken from the table

Preparation: There will be two teams, each consisting of one Plotter and one Coordinate Maker. One team will look for a relationship that has a positive correlation. The other team will look for a relationship that has a negative correlation. The data will come from the table below, which shows how people in different age groups spend their leisure time. All quantities represent millions of people.

Age	Movies	Sports events	Amusement	Exercise	Play sports	Gardening
25–34	70	47	68	67	59	51
35–44	68	43	58	67	52	57
45–54	58	35	44	62	44	64
55–64	50	23	30	62	34	63

Procedures:

1. Each member chooses a role.

2. Each team discusses the data and chooses an event that they think might be positively or negatively correlated with age.

   _____

3. The Coordinate Maker writes down four coordinate pairs by determining a way to represent the age group as an *x*-coordinate. The Plotter makes a scatter plot of the data.

   _____

4. Each team decides whether the graph shows a positive or a negative correlation.

   _____

5. Each team finds an equation for the least-squares line for their data and graphs it on the graphics calculator.

   _____

6. Teams share their work and discuss the relationships that they have found among the data in the table.

   _____

Algebra 2   Cooperative-Learning Activity   5

NAME _____ CLASS _____ DATE _____

Cooperative-Learning Activity
1.6 Solving Equations

Group members: 4

Materials: timer

Roles: Solver   solves equations

Checker   checks solutions using substitution

Preparation: Your group will form two teams, Team A and Team B, each consisting of a Solver and a Checker. Each team will try to be the first to finish solving and checking equations. In order to be the winning team, all solutions must be correct.

Procedures:
1. Each member chooses a role.

2. The Solvers in Teams A and B solve each equation on their own paper. The corresponding Checker checks the solutions. The Checker writes the solution to each equation and the amount of time it took to solve, check, and revise the solution in the blanks below. The Solver may not move on to the next equation until the Checker is satisfied that the solution is correct.

   a. $4x - 3 = 13$ Solution _____   Solving time _____

   b. $9\left(\frac{2}{3}x - \frac{1}{3}\right) = 12$ Solution _____   Solving time _____

   c. $12x = 5 - 3x$ Solution _____   Solving time _____

   d. $\frac{14 - x}{2} = 6 - x$ Solution _____   Solving time _____

   Total solving time, Team A _____

   Total solving time, Team B _____

3. The team members switch roles. Then they solve and check the equations below.

   a. $-13 = -\frac{5}{7}x - 5$ Solution _____   Solving time _____

   b. $\frac{3(x - 5)}{4} = x + 6$ Solution _____   Solving time _____

   c. $\frac{2}{5}x + \frac{1}{10} = -5$ Solution _____   Solving time _____

   d. $\frac{1}{4}x - 1 = \frac{1}{5}x + 3$ Solution _____   Solving time _____

   Total solving time, Team A _____

   Total solving time, Team B _____

NAME _____ CLASS _____ DATE _____

Cooperative-Learning Activity
1.7 Using Inequalities to Explore Allocation of Income

Group members: 2

Materials: no special materials

Roles: **Inequality Writer** writes an inequality to represent a situation
Inequality Solver solves an inequality to answer a related question

Preparation: Mr. and Ms. Clarkson have three children. The parents are trying to allocate money on a monthly basis for the children's future needs. They are also planning a family vacation. Finally, they are anticipating future housing needs.

Listed below are questions that Mr. and Ms. Clarkson are trying to answer.

- They set aside $300 monthly ($100 for each child) into a fund for the future education of each child. They want to increase that monthly allocation. If the Clarksons can afford to set aside at most $570 monthly, how much more money should they set aside for each child?

- The Clarksons are planning to take a long-awaited family vacation. They can afford to set aside at least $3000, but no more than $3600 for the trip. Their expenses include travel expenses and hotel accomodations totaling $1800. Once at their destination, they can afford at least $800, but no more than $1200 on entertainment and meals. How much money will they have left for additional expenses?

- The Clarksons believe that next year they will need to build an addition to their home. Two contractors estimate that the cost will be between $30,000 and $40,000. One contractor estimates the cost of materials to be between $12,000 and $16,000. The second contractor estimates the cost of materials to be between $15,000 and $18,000. What labor costs can the Clarksons expect to pay?

Procedures: 1. Each member chooses a role.

2. The Inequality writer sets up an inequality in one variable to answer the first of the Clarkson's questions. _____

3. The Inequality Solver solves the inequality. _____

4. Both members act together as Inequality Writer and Inequality Solver to answer the second question.

_____

5. Both members act together as Inequality Writer and Inequality Solver to answer the third question.

_____

Algebra 2   Cooperative-Learning Activity   7

NAME _____ CLASS _____ DATE _____

Cooperative-Learning Activity
1.8 Absolute-Value Inequalities

Group members: 3

Materials: no special materials

Responsibilities: Solve the inequality for one row and graph these solutions on the number line.

Preparation: You will solve and graph inequalities on the three number lines labeled *top*, *middle*, and *bottom*.

Procedures:
1. Decide who will solve the absolute-value inequalities in each part of Procedure 2.

2. Each member solves the inequalities for his or her row and graphs them on the appropriate number line below. The person working with the bottom row also checks the work of the other members.

	Top	Middle	Bottom
a.	$\|x + 9\| \leq 1$	$\|x + 3\| \leq 1$	$\|x + 6\| \leq 1$
b.	$\|6 - x\| \leq 1$	$\left\|x + \frac{1}{4}\right\| \leq \frac{3}{4}$	$\|6 - 2x\| \leq 2$
c.	$\|-3x + 9\| \leq 3$	$\|x - 7.5\| \geq 0.5$ and $5 \leq x \leq 10$	$\|15 + 5x\| \leq 5$
d.	$\|x + 0.5\| \geq 0.5$ and $-4 \leq x \leq 1$	$\|2x + 18\| \leq 2$	$\|3x\| \leq 3$
e.	$\|-2x + 18\| \leq 2$	$\|2x - 6\| \leq 2$	

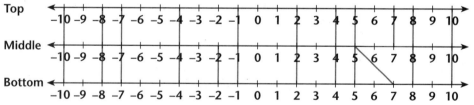

3. What appears on the three number lines above? Discuss any methods that you found helpful in solving absolute-value inequalities.

_____

_____

_____

NAME _____ CLASS _____ DATE _____

Cooperative-Learning Activity
2.1 Classifying Numbers

Group members: 3

Materials: index card

Roles: **Number Generator** generates numerical expressions

Classifier classifies numbers by using a Venn diagram

Checker checks accuracy of Venn diagram

Preparation: Enlarge the diagram at right.

The Venn diagram at right shows four sets of numbers: multiples of 2, multiples of 3, multiples of 5, and those numbers that are not multiples of 2, 3, or 5. Your group will identify a given number as a member of one of these sets.

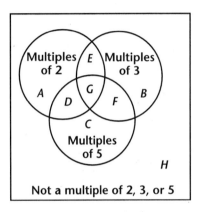

Procedures:
1. Each member chooses a role.

2. The Number Generator writes ten numerical expressions on an index card. Each of the expressions belongs to one of the sets shown in the Venn diagram above. However, none of the numbers can immediately be identified as a member of any of the sets. For example, the Number Generator may write $\frac{10 + 20}{3}$.

3. The Classifier simplifies the numbers from the index card. Then the Classifier writes each number in the appropriate region. For example, for $\frac{10 + 20}{3}$, the Classifier writes 10 in region D.

4. The Checker checks the Classifier's work.

5. As a group, discuss each of the following questions:

What must be true of a number that belongs in region G?

_____

What can you say about a positive integer whose units digit is 0?

_____

What can you say about a positive integer whose units digit is 3?

_____

Algebra 2   Cooperative-Learning Activity   9

NAME _____ CLASS _____ DATE _____

Cooperative-Learning Activity
2.2 Equivalent Powers

Group members: 2

Materials: grid paper

Responsibilities: You will identify equivalent expressions.

Preparation: The table contains 4 expressions written in many different forms.

$\left(\dfrac{5x^0y^5}{25^{\frac{1}{2}}y^5}\right)$	$\left(-\dfrac{16}{64}\right)^0$	1	$\left(\dfrac{-8(-x)^5}{y^{-8}}\right)^{-1}$	$2^{\frac{2}{3}}$
$(8x^9y^{15})^{\frac{1}{3}}$	$\dfrac{4^{-1}x^3y^{-5}}{2x^8y^3}$	$(-4^{\frac{1}{3}})^0$	$\sqrt[3]{4}$	$\dfrac{(2x^5y^{10})^{-3}}{x^{-7}y^{-22}}$
$(2^0)^{\frac{1}{9}}$	$2x^3y^5$	$\left(\dfrac{64}{16}\right)^{\frac{1}{3}}$	$(2x^5y^8)^0$	$\left(\dfrac{2^3x^{10}y^{16}}{x^5y^8}\right)^{-1}$
$(4^0)^{\frac{2}{6}}$	$\dfrac{4^{\frac{1}{2}}xy^{-2}}{8^0x^{-2}y^{-7}}$	$(2^{\frac{1}{3}})^2$	$\dfrac{3^0x^{-5}y^{-2}}{2^{-1}x^{-8}y^{-7}}$	$\dfrac{1}{8x^5y^8}$

Procedures:

1. Divide up the expressions in the table above. Together, simplify each expression using positive exponents only.

2. Write four equations that show which expressions are equivalent to one another. Two expressions are considered equivalent if they have the same simplification.

_____

_____

_____

_____

3. Discuss the importance of simplifying an expression before using it in further work.

_____

_____

NAME _____ CLASS _____ DATE _____

Cooperative-Learning Activity
2.3 Using Functions

Group members: 4

Materials: pencil, grid paper, calculator

Responsibilities: Use a function to represent a real-world situation.

Preparation: You buy a new car for $32,000. You make a 10% down payment and take out a loan for the remaining amount. One option is a loan is for 4 years at the annual interest rate of 8%. At another dealership, a loan is for 3 years at the annual interest rate of 10%. The formula used to calculate the monthly payment is shown below.

$$M = \frac{Pr(1 + r)^n}{(1 + r)^n - 1} \quad \begin{cases} M: \text{monthly payment} \\ P: \text{amount of loan} \\ r: \textbf{monthly } \text{interest rate} \\ n: \text{number of payment periods} \end{cases}$$

Procedures:

1. Divide into two teams.

2. Each team completes the table below. (Remember that the amount of the loan is not the total cost of the automobile.)

Option 1: 4 years at 8%			Option 2: 3 years at 10%		
Number of payments	Monthly rate	Monthly payment	Number of payments	Monthly rate	Monthly payment

3. Let x represent the number of monthly payments made. Each team completes the table below.

Option 1: 4 years at 8%		Option 2: 3 years at 10%	
x	Balance due	x	Balance due
1		1	
2		2	
3		3	
4		4	

4. Each team writes a function for the balance due in terms of x and their respective monthly payments.

_____

Algebra 2   Cooperative-Learning Activity   **11**

NAME _____ CLASS _____ DATE _____

Cooperative-Learning Activity
2.4 Function Machines

Group members: 4

Materials: pencil, 3 index cards

Roles: *f*-machine acts as function *f*
   g-machine acts as function *g*
   h-machine acts as function *h*
   Director reads the expression and directs the Function Machines

Preparation: The Function Machines will use the functions shown below.

$$f(x) = x^2 - 49 \qquad g(x) = x + 7 \qquad h(x) = 20x$$

Procedures:

1. Each member chooses a role. Each Function Machine writes his or her function on an index card.

2. The Director chooses an integer value for *x*. The Function Machines evaluate their functions for *x* and then work together to evaluate the functions below.

   a. $f(x) + g(x)$ _____   **b.** $g(x) - h(x)$ _____

   c. $(gh)(x)$ _____   **d.** $(hg)(x)$ _____

   e. $\dfrac{f}{h}(x)$ _____   **f.** $\dfrac{f}{g}(x)$ _____

3. Switch roles and repeat Procedure 2, but now the Function Machines use the variable *x*.

   a. $f(x) + g(x)$ _____   **b.** $g(x) - h(x)$ _____

   c. $(gh)(x)$ _____   **d.** $(hg)(x)$ _____

   e. $\dfrac{f}{h}(x)$ _____   **f.** $\dfrac{f}{g}(x)$ _____

4. Switch roles again so that there is a new Director and each member has a different function than before. The director directs the composition of functions below and may choose any value for *x*.

   a. $(f \circ g)(x)$ _____   **b.** $(g \circ f)(x)$ _____

   c. $(h \circ g)(x)$ _____   **d.** $(g \circ h)(x)$ _____

5. Which operations can you perform in any order?

   _____

   _____

12   Cooperative-Learning Activity   Algebra 2

NAME _____ CLASS _____ DATE _____

Cooperative-Learning Activity
2.5 Interpreting Inverses

Group members: 2

Materials: no special materials

Roles: Function Maker   chooses variables and writes an equation expressing the relationship

Inverse Maker   writes a new equation showing the inverse, if possible

Preparation: A. A swimming pool has a length of 20 feet and a width of 40 feet. The volume of water needed to fill the pool is a function of the desired depth.
B. The amount of material required to make a square table cloth depends on the width of the table cloth.
C. For some phone companies, the phone charge is a fixed charge plus a charge per minute.
D. A local department store is clearing out their merchandise by offering a discount of 30% on a price that was already marked down 30%.

Procedures: 1. Each member chooses a role.

2. **a.** The Function Maker writes an equation for situation A.

   _____

   b. The Inverse Maker writes an equation for the inverse relationship.

   _____

   c. Interpret the inverse relationship, and decide whether this relationship is a function.

   _____

3. Repeat Procedure 2 for situation B.

   a. _____

   b. _____

   c. _____

4. Switch roles and repeat Procedure 2 for situation C or situation D.

   a. _____

   b. _____

   c. _____

Algebra 2   Cooperative-Learning Activity   **13**

NAME _____ CLASS _____ DATE _____

Cooperative-Learning Activity
2.6 The Rounding Function

Group members: 3

Materials: no special materials

Responsibilities: Graph and create a function that rounds real numbers.

Preparation: Your group will work together to graph and classify the rounding function.

Procedures:
1. A decimal number is rounded down to the nearest interger when its decimal part is less than one-half and rounded up when its decimal part is greater than or equal to one-half. Write 5 ordered pairs whose x-coordinate is between -1 and 0 and whose y-coordinate is the result of rounding the x-coordinate to the nearest integer.

   _____

2. Plot the points on the grid at right below.

3. Plot five more points between -1 and 0.

4. Write a formula for this function for all x-coordinates between -1 and 0.

   _____

5. Repeat Procedures 1–3 for the intervals between 0 and 1, between -1 and -2, and between 1 and 2.

6. Write a formula for the function you have plotted.

   _____

7. Can the rounding function be classified as one of the special functions in this lesson? Which one?

   _____
   _____
   _____
   _____
   _____
   _____
   _____

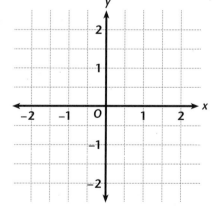

14 Cooperative-Learning Activity   Algebra 2

NAME _____ CLASS _____ DATE _____

Cooperative-Learning Activity
2.7 Families of Functions

Group members: 4

Materials: tracing paper

Roles: Absolute Value Function models $y = |x|$
Square Function models $y = x^2$
Linear Function models $y = x$
Leader directs the functions in transformations

Preparation: The members representing functions will write new functions based on directions given by the Leader.

Procedures:
1. Each member chooses a role.

2. Each member representing a function chooses the appropriate graph shown below and reproduces it on tracing paper. The tracing should include the graph, the axes, and tick marks indicating the locations of the integers on the axes.

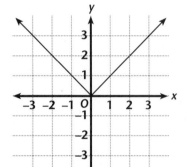
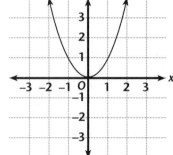
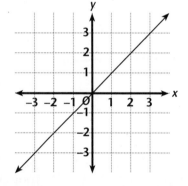

3. The Leader gives the first set of instructions listed below. Each other member sketches a new graph and writes an equation. The Functions go back to the original graph after each instruction.
   a. Move the graph 4 units to the left.
   b. Move the graph 8 units to the right and 6 units down.
   c. Move the graph 2 units to the left and 5 units up.

4. Switch roles. The new Leader reads the instructions in Procedure 2. This time, the other members do not go back to the original graph after each instruction.

5. Discuss how to represent a collection of transformations as a single transformation of the graph of the parent function.

_____

_____

Algebra 2 — Cooperative-Learning Activity — 15

NAME _____ CLASS _____ DATE _____

Cooperative-Learning Activity
3.1 Modeling and Solving a Coin Problem

Group members: 3

Materials: no special materials

Roles: **Equation Writer** writes an equation to model a coin problem

Grapher graphs the equation provided by the Equation Writer

Solver finds integer solutions to the equation

Preparation: Suppose that you have $0.75 in nickels and dimes. How many coins could you have in all? To begin, let n represent the number of nickels and let d represent the number of dimes.

Procedures: 1. Each member chooses a role.

2. The Equation Writer writes an equation to represent the situation above. The equation should be written in standard form.

_____

3. The Grapher solves the equation for d.

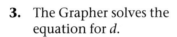

The Grapher uses the coordinate grid at right to graph the equation.

4. The Solver circles points on the graph where n and d are both whole numbers. The Grapher uses substitution to check that these points are actually located on the graph.

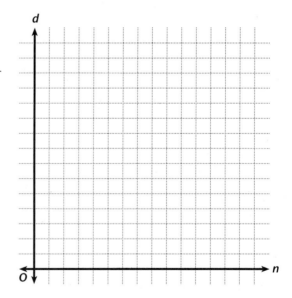

5. As a group, find four possible answers to the question "How many coins could you have in all?"

_____

_____

6. The Equation Writer writes four systems of equations, one for each of the possible answers in Procedure 5.

_____

16 Cooperative-Learning Activity   Algebra 2

NAME _____ CLASS _____ DATE _____

Cooperative-Learning Activity
3.2 Making Systems From Solutions

Group members: 4

Materials: four number cubes

Responsibilities: Write and solve systems of equations.

Preparation: Teams will generate nine systems of equations. Teams will exchange systems and solve.

Procedures:
1. Divide into two teams. Each team has two number cubes.

2. In the space below, each team generates three sets of systems of equations for a total of nine systems. Each set of systems of equations is generated as follows:

   a. Roll the number cubes. These numbers represent the x-coordinate and y-coordinate of the solution of the first system. Write a system of two equations whose graphs contain the ordered pair. The x- and y-coefficients must be chosen so that the system is consistent and independent.

   b. Alter one of the equations so that the system is inconsistent.

   c. Alter one of the equations from the original system so that the system is dependent.

3. Each team rewrites the systems in a different order for the other team.

4. Teams exchange systems and solve using any method.

5. What are some of the characteristics for inconsistent systems and dependent systems?

   _____

   _____

Algebra 2

NAME _____ CLASS _____ DATE _____

Cooperative-Learning Activity
3.3 Coin Inequalities

Group members: 2

Materials: no special materials

Roles: **Solver** graphs an inequality in two variables and its solution
Solution Counter counts solutions with whole number coordinates

Preparation: Suppose that you have $0.50 in nickels and dimes. If n represents the number of nickels and d represents the number of dimes, then $0.05n + 0.10d = 0.50$ represents your total amount of money. The diagram at right shows the graph of $0.05n + 0.10d = 0.50$. The axes on the grid represent nickels and dimes. The money value of a point can be found by finding the worth of the nickels and dimes represented by (n, d). You will use these ideas to investigate solutions to an inequality.

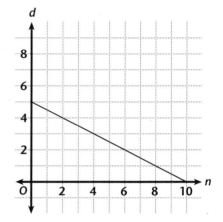

Procedures:
1. Each member chooses a role.

2. The Solver shades the region in the plane that is the solution to the inequality $0.05n + 0.10d \leq 0.50$.

3. The Solution Counter determines the number of points whose coordinates are whole numbers that solve the inequality. _____

4. For each graph below, work together to count the number of nickels and dimes that satisfy a nickel/dime inequality. They write the inequality represented by each graph.

_____

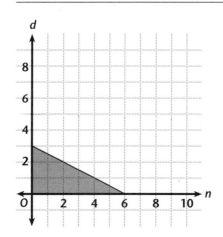   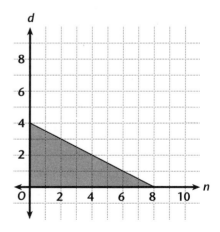

18 Cooperative-Learning Activity Algebra 2

NAME _____ CLASS _____ DATE _____

Cooperative-Learning Activity
3.4 Solving a System of Inequalities

Group members: 2

Materials: two differently-colored pencils

Responsibilities: Model an inequality problem, and interpret the graph and solutions of a system of inequalities.

Preparation: Promoters at a rock concert are selling T-shirts with the artist's portrait and CDs with the concert music. The promoters expect that they will sell at least twice as many CDs as T-shirts. They will sell the T-shirts for $15 and the CDs for $18. Their goal is to make at least $270.

Procedures:

1. Choose variables for the problem.

   _____

2. Discuss the situation and write two inequalities, one involving the numbers of items sold and the other involving the goal for revenue. Each group member should write one of the inequalities.

   _____

3. Graph the boundary lines for the inequalities created in Procedure 2 on the grid at right, using different colors.

4. Discuss which region should be shaded for each inequality and shade the region that represents the solution of the system.

   _____

5. What is the minimum number of T-shirts and CDs the promoters should have available?

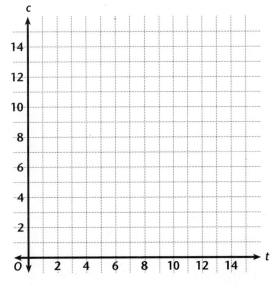

   _____

6. Discuss whether the promoters have enough information to determine the maximum number of T-shirts and CDs that they should bring to the concert. Explain why or why not.

   _____

   _____

Algebra 2   Cooperative-Learning Activity   19

NAME _____ CLASS _____ DATE _____

Cooperative-Learning Activity
3.5 Evaluating the Objective Function

Group members: 3

Materials: no special materials

Roles: **Inside Evaluator** evaluates the objective function for 5 points inside the feasible region

Border Evaluator evaluates the objective function for 5 points on the boundary of the shaded region (but not at vertices), selecting at least one point from each border

Vertex Evaluator evaluates the objective function at the vertices of the feasible region

Preparation: The feasible region for a linear programming problem is the shaded region at right.

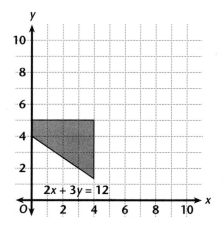

Procedures:

1. Each member chooses a role.

2. Evaluate the objective function $P = 2x - 3y$ for the coordinates as assigned by the description of roles. Complete the tables below. Devise strategies to find points that give the maximum and minimum values of the objective function.

Inside	
Coordinates	P

Borders	
Coordinates	P

Vertices	
Coordinates	P

3. Discuss your strategies and the values for *P* that appear to be the maximum and minimum. Decide where the maximum and minimum values of the objective function can be found.

NAME _____ CLASS _____ DATE _____

Cooperative-Learning Activity
3.6 Modeling Time/Distance Travel

Group members: 2

Materials: number cube

Roles: **Timer** rolls number cube to determine time

Traveler determines location after walking for time called out by timer.

Preparation: A map is shown below. The situations below the map give directions for the Traveler.

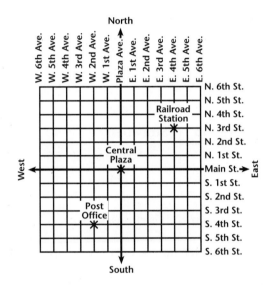

Situation A: Start at the corner of E. 6th Ave and S. 6th St.
Each minute, go 2 blocks west and 1 block north.

Situation B: Start at the corner of W. 6th Ave and N. 6th St.
Each minute, go 2 blocks east and 2 blocks south.

Procedures:
1. The Timer rolls the number cube to obtain the number of minutes, n. The Timer then says to the Traveler, "n minutes have passed. Where are you?" For example, if a 2 is rolled on the number cube, the Timer says, "Two minutes have passed. Where are you?"

2. The Traveler gives the intersection where he or she is now located, according to the directions given in Situation A.

3. Repeat Procedures 1 and 2.

4. Switch roles and repeat the entire process, using Situation B.

5. Write parametric equations to represent Situation A and Situation B in terms of time, t.

Situation A: _____   Situation B: _____

Algebra 2   Cooperative-Learning Activity   21

NAME _____ CLASS _____ DATE _____

Cooperative-Learning Activity

4.1 Handling Store Inventories With Matrices

Group members: 3

Materials: no special materials

Roles: **Store Manager** gives detailed instructions to the Inventory Accountant

Inventory Accountant writes matrices from data and follows instructions given by the Store Manager

Auditor checks the calculations of the Inventory Accountant

Preparation: The tables below show the ice cream inventory (in cases) in the 3 stores of a chain at the beginning of June.

Store A

	Regular	Lite
Chocolate	48	16
Strawberry	30	16
Vanilla	50	40
Mocha	30	16

Store B

	Regular	Lite
Chocolate	60	25
Strawberry	40	30
Vanilla	50	40
Mocha	15	30

Store C

	Regular	Lite
Chocolate	45	30
Strawberry	20	15
Vanilla	15	10
Mocha	16	10

Procedures:

1. Each member chooses a role.

2. On a separate sheet of paper, the Inventory Accountant writes three 4×2 matrices, A, B, and C. Each of these matrices should represent the inventory at the corresponding store.

3. The Store Manager asks for a matrix representing the total inventory for the chain by telling the Inventory Accountant which entries to add. For example, the total number of cases of regular chocolate ice cream is given by $a_{11} + b_{11} + c_{11}$. The Auditor checks the calculations.

4. Switch roles. The Store Manager is planning to have a large promotional event at store C, so he or she plans to transfer half of the inventory at store A to store C. The Inventory Accountant's first task is to find a matrix H representing one-half of the inventory at store A. What is H?

5. Let A', B', and C' be matrices representing the inventory of each store after the transfer takes place. The Store Manager writes equations showing how to find A', B', and C' in terms of A, B, C, and H.

6. The Inventory Accountant finds the new matrices A', B', and C'. The Auditor checks the calculations.

7. The Store Manager finds a new matrix representing the chain's total inventory after the transfer. How does this matrix compare with the matrix in Procedure 3? As a group, explain why this result makes sense.

22 Cooperative-Learning Activity   Algebra 2

NAME _____ CLASS _____ DATE _____

Cooperative-Learning Activity
4.2 Pricing Theater Seats

Group members: 2

Materials: no special materials

Responsibilities: Use matrix multiplication to solve the problem below.

Preparation: A community theater group uses a theater that holds 500 people. The theater has 25 rows. The group is to come up with a plan to raise $8000 by dividing the theater into 3 sections where the prices are as follows:

Section 1	$20
Section 2	$15
Section 3	$10

Procedures:

1. Write a 3 × 1 matrix, A, containing the 3 prices.

$$A = \begin{bmatrix} \\ \\ \end{bmatrix}$$

   Write a 1 × 3 matrix containing possible numbers of seats for each price range. All seats in a row must be sold at the same price. Assume that the theater will be filled.

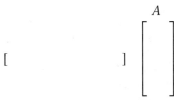

2. Use matrix multiplication to find the total income.

   _____

3. Try different values for the number of seats in each section until you find values that give a total income of $8000.

   _____

4. Write a summary of your strategies for finding the solution.

   _____
   _____
   _____

Algebra 2   Cooperative-Learning Activity

NAME _____ CLASS _____ DATE _____

Cooperative-Learning Activity
4.3 Exchanging Secret Questions

Group members: 2

Materials: calculator with matrix capability

Responsibilities: Establish an assignment table, and then correspond by encoding and decoding messages.

Preparation: You will work together to come up with an assignment table different than the one in Lesson 4.3 of the textbook. Then you will encode and decode messages. The matrix to use for coding will be $\begin{bmatrix} 5 & -6 \\ -6 & 5 \end{bmatrix}$.

Procedures:

1. Work together to establish an assignment table that is different from the one given in Lesson 4.3.

—	G	N	U	
A	H	O	V	
B	I	P	W	
C	J	Q	X	
D	K	R	Y	
E	L	S	Z	
F	M	T	?	

2. First, each member acts as an encoder. Each member thinks of a question to ask the other member. He or she then uses the coding matrix to encode the message.

3. Find the inverse of the coding matrix independently and check to see if your matrices are the same. If they are not the same, decide which one is correct.

4. Each member now becomes a decoder and decodes the message.

5. The members encode responses to the questions and exchange matrices again.

6. The members decode the answers.

NAME _____ CLASS _____ DATE _____

Cooperative-Learning Activity
4.4 Preparing Trail Mix

Group members: 2

Materials: graphics calculator

Roles: **Calculator Solver** solves system by using graphics calculator and inverse matrices

Hand Solver solves system without calculator

Preparation: A wholesale health food store owner sometimes gets special requests for trail mixes made of raisins and sunflower seeds. The current stock consists of the following trail mixes.

Trail mix 1
55% sunflower seeds

Trail mix 1
20% sunflower seeds

Procedures:

1. Each member chooses a role.

2. A retailer orders 20 pounds of mix that is 45% sunflower seed. Complete the table below to help set up a system of equations that will determine how many pounds of each mix must be combined.

	Trail Mix 1	Trail Mix 2	Mixture
Number of pounds			
Amount of sunflower seeds			

3. Write a system of equations in two variables to answer the question.

   _____

4. The Hand Solver solves the system by using the substitution or elimination method. The Calculator Solver solves the system by using a graphics calculator, rounding solutions to nearest tenth. Compare solutions.

   _____

5. Another retailer orders 20 pounds of trail mix that is 25% sunflower seeds. Write a new system, switch roles, and solve.

   _____

   _____

6. How can you approximate the amount needed for 20 pounds of a 50% mixture without using a system of equations?

   _____

Algebra 2

NAME _____ CLASS _____ DATE _____

Cooperative-Learning Activity
4.5 Working Together to Find the Row-Reduced Matrix

Group members: 3

Materials: no special materials

Roles: **Director** writes elementary row operations
Row Operator performs elementary row operations
Checker checks Row Operator's work

Preparation: Your group will find the reduced row-echelon form of each matrix.

$$A = \begin{bmatrix} 3 & 0 & -1 & 4 \\ -2 & 3 & 1 & 6 \\ 1 & -1 & 0 & 5 \end{bmatrix} \quad B = \begin{bmatrix} 1 & -1 & 3 & 2 \\ -2 & 2 & -6 & -1 \\ 4 & 1 & -1 & 2 \end{bmatrix}$$

Procedures: 1. Each member chooses a role. The members will find a reduced row-echelon form for Matrix A.

2. The Director writes down the first row operation using row operation notation. The Row Operator performs the operation. The Checker makes sure that the work is done properly.

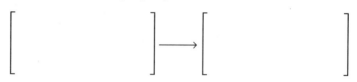

3. Repeat Procedure 2 until you find the matrix in reduced row-echelon form.

4. Switch roles and find the reduced row-echelon form of Matrix B.

5. Discuss how Matrix B is different from Matrix A. If Matrix B represented a system of linear equations, what would the system be? What does the reduced row-echelon form of B tell you about the system?

NAME _____ CLASS _____ DATE _____

Cooperative-Learning Activity
5.1 Multiplying Binomials Using Area Models

Group members: 3

Materials: no special materials

Roles: **Rectangle Maker** draws a rectangle whose side lengths are binomials

Area Finder finds area by adding the parts of the rectangle

Evaluator makes a table and evaluates for different values of x

Preparation: Your group will find the products of binomials by drawing and finding the areas of rectangles whose side lengths represent the binomials. For example, the rectangle at right represents $(x + 3)(x + 9)$. The area of this rectangle can be found by adding the areas of the four interior rectangles. Thus, $(x + 3)(x + 9) = x^2 + 9x + 3x + 27 = x^2 + 12x + 27$. Then you will calculate the area for different values of x.

Procedures:

1. Each member chooses a role.

2. The Rectangle Maker draws a rectangle similar to the one above, whose sides are given by $x + 3$ and $x + 5$.

3. The Area Finder finds the area of the rectangle by adding the areas of the interior rectangles. _____

4. The Evaluator chooses 5 positive values for x and completes the table.

x	$(x + 3)(x + 5)$	Area

5. As a group, find the minimum value of the area.

_____

6. Switch roles and repeat Procedures 2–5 for the rectangle whose area is modeled by $(2x + 1)(x + 3)$.

x	$(2x + 1)(x + 3)$	Area

_____

7. Describe how to find the area for a rectangle with side lengths of $ax + b$ and $x + c$.

_____

_____

Algebra 2   Cooperative-Learning Activity   27

NAME _____ CLASS _____ DATE _____

Cooperative-Learning Activity

5.2 Finding Perimeter and Area

Group members: 2

Materials: no special materials

Responsibilities: You will work together to solve the following problems.

Preparation: The diagram at the bottom of the page shows how quadrilateral *WXYZ* fits snugly inside rectangle *ABCD*. The dashed lines indicate the relationships between points *X* and *Z* and points *W* and *Y*. The diagram also shows four right triangles that will help solve the problems below.

Procedures:

1. Find the following lengths:

   BW = _____   BX = _____   YD = _____   DZ = _____

2. Use the given information and answers from Procedure 1 to find:

   WX = _____   YX = _____   ZD = _____   WZ = _____

3. Compute the perimeter of quadrilateral *WXYZ*. _____

4. Working together, devise and use a strategy to find the area of quadrilateral *WXYZ*, using the area formulas for a triangle and rectangle.   _____

   Is the Pythagorean Theorem needed to find the area? Explain.

   _____

   _____

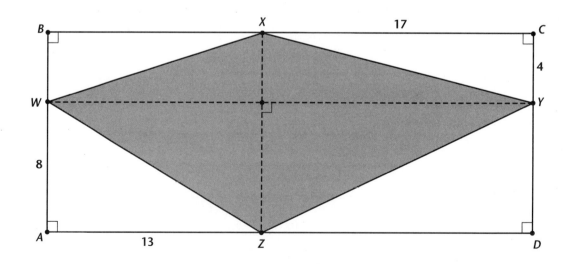

28   Cooperative-Learning Activity   Algebra 2

NAME _____ CLASS _____ DATE _____

Cooperative-Learning Activity
5.3 Finding Number Pairs

Group members: 4

Materials: no special materials

Roles: **Addends Finder** finds pairs of integers whose sum is a given number. The pairs should include some where the two addends have different signs.

Factors Finder finds all pairs of integers whose product is a different given number

Preparation: Use the following list of sums and products.

	Sum	Product
A	−10	21
B	−2	−48
C	1	−72
D	−5	4
E	−12	27

	Sum	Product
F	−6	8
G	10	−24
H	13	−30
I	−14	40
J	−9	−36

Procedures:
1. Form two teams, each team with an Addends Finder and a Factors Finder.

2. Working independently, the Addends Finders write pairs of numbers that add to the sum in rows A–J while the Factors Finders write pairs of numbers that multiply to the product in rows A–J.

3. The members of each team compare their lists and find the number pair that is common to each list. The team that finishes first is the winner.

4. The teams make up a similar table with 5 problems and exchange tables.

Sum	Product

5. The Addends Finders and Factors Finders switch roles and write number pairs according to Procedure 2.

Algebra 2   Cooperative-Learning Activity   29

NAME _____ CLASS _____ DATE _____

Cooperative-Learning Activity
5.4 Organizing A Solution Process

Group members: 2

Materials: no special materials

Responsibilities: Make an organized plan for solving equations of the form $ax^2 + bx + c = 0$ by completing the square.

Preparation: Any equation of the form $ax^2 + bx + c = 0$, where $a \neq 0$, can be written equivalently as $a(x - h)^2 + k = 0$ for some real numbers h and k.

Procedures:

1. From the equations below, members take turns choosing equations until each member has chosen three equations.

   A. $x^2 + 5x + 6 = 0$ B. $x^2 - 7x + 10 = 0$ C. $x^2 + 11x + 30 = 0$

   D. $2x^2 + 7x - 15 = 0$ E. $3x^2 + 19x - 14 = 0$ F. $6x^2 - 7x - 5 = 0$

2. Each member writes his or her equations in the form $a(x - h)^2 + k = 0$.

   A. _____   B. _____   C. _____

   D. _____   E. _____   F. _____

3. Members solve their equations from Procedure 2 and write the solutions.

   A. _____   B. _____   C. _____

   D. _____   E. _____   F. _____

4. Working together, write a two-step plan that you can follow to solve an equation of the form $ax^2 + bx + c = 0$ that includes completing the square.

   I. _____

   II. _____

5. Solve $a(x - h)^2 + k = 0$ to write a formula that will give x. Write each step in the solution, and give a reason for each step.

   Step   Reason

   _____   _____

   _____   _____

   _____   _____

   _____   _____

30   Cooperative-Learning Activity   Algebra 2

NAME _____ CLASS _____ DATE _____

Cooperative-Learning Activity
5.5 Using the Quadratic Formula to Plan a Patio

Group members: 3

Materials: no special materials

Roles: Architect   draws and labels sketches
Engineer   writes relevant equations
Contractor   solves equations

Preparation: A homeowner wants to build a patio around a 35-foot by 20-foot rectangular swimming pool. The homeowner has two options and a budget of $40,200 for the patio.

Option 1   Surround the pool with a concrete band of equal width on all sides. The concrete costs $50 per square foot.

Option 2   Surround the pool with a tile band of equal width on all sides. The tile costs $80 for each square foot tile.

Procedures:
1. Each member chooses a role.

2. The Architect draws a sketch of the pool and patio while the Contractor calculates the square footage that can be purchased for each material, assuming that the entire budgeted amount is used.

3. The Engineer writes equations to represent Options 1 and 2.

   Option 1 _____   Option 2 _____

4. The Contractor solves the two equations by using the quadratic formula.

   Option 1   Option 2

   _____   _____

   _____   _____

   _____   _____

   _____   _____

5. Discuss the two options and make a recommendation.

   _____

   _____

Algebra 2   Cooperative-Learning Activity   **31**

NAME _____ CLASS _____ DATE _____

Cooperative-Learning Activity
5.6 Geometry in Complex Number Addition

Group members: 2

Materials: no special materials

Responsibilities: Work together to discover a geometric relationship between two points that represent complex numbers and the point that represents their sum.

Preparation: Refresh your memory of the complex plane, the real and imaginary axes, and how to graph a complex number in the complex plane.

Procedures:
1. One member finds the sum of the two complex numbers in part **a** and in part **b**. Then he or she writes and graphs the sum. The other member does the same thing with parts **c** and **d**. Each member joins the two given points with the point that represents the sum to form a quadrilateral.

a. sum: _____  b. sum: _____

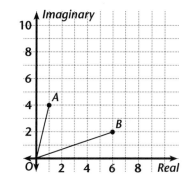 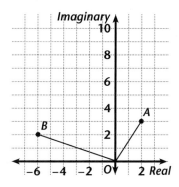

c. sum: _____ d. sum: _____

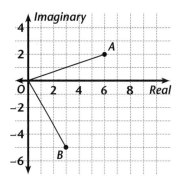 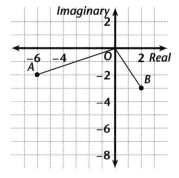

2. Working together, write a conjecture about the quadrilaterals that you have sketched.

_____

3. Write a rule that tells how to find the sum of two complex numbers geometrically.

_____

NAME _____ CLASS _____ DATE _____

Cooperative-Learning Activity
5.7 Exploring Second Differences and Quadratic Models

Group members: 2

Materials: no special materials

Responsibilities: Draw inferences about first, second, and third differences in quadratic number patterns.

Preparation: If a number pattern is linear, then the first differences are constant.

Procedures:
1. One member fills in each row of differences for the function tables in part **a** and part **b**. The second member does the same thing for the tables in part **c** and part **d**.

   a.

n	0	1	2	3
$P(n)$	2	1	-2	-7

   b.

n	1	2	3	4
$P(n)$	-3	3	13	27

   c.

n	0	1	2	3
$P(n)$	7	5	-1	-11

   d.

n	1	2	3	4
$P(n)$	4	8	14	22

2. Based on work done in Procedure 1, write a statement about first, second, and third differences for quadratic number patterns.

   first differences _____

   second differences _____

   third differences _____

3. In Procedures 1 and 2, you assumed that the tables represented quadratic functions. To verify this assumption, one member finds a quadratic model for the tables in parts **a** and **c**. The other member finds a quadratic model for the tables in parts **b** and **d**.

   a. _____   b. _____

   c. _____   d. _____

Algebra 2

NAME _____ CLASS _____ DATE _____

Cooperative-Learning Activity
5.8 Analyzing Profits

Group members: 3

Materials: graphics calculator

Roles: **Cost Analyst** writes cost function, C

Revenue Analyst writes equation for revenue function, R

Profit Analyst writes equation for profit function, P

Preparation: A group of students produces and sells ceramic vases. The cost of making the vases includes a fixed cost of $75 for the use of a potter's wheel, plus $20 for materials and labor for each vase.

An order must be at least 10 vases, which are priced at $30 per vase. For orders of more than 10 vases, the price per vase is reduced by one dollar for each vase over the minimum order.

Procedures: 1. Each member chooses a role.

2. The Cost Analyst and Revenue Analyst write the cost and revenue functions described above in terms of x, the number of vases over the minimum order.

   $C(x) =$ _____ $R(x) =$ _____

3. The Profit Analyst graphs the functions $y = C(x)$ and $y = R(x)$ on the calculator and sketches those graphs on the grid at right. The Profit Analyst writes the profit function $P(x) = R(x) - C(x)$ and adds that function to the graph.

4. Discuss the graphs of the three functions and how they are related. Then write a system of inequalities to describe the region of the graph where the students will make a profit. The Profit Analyst determines at what point the profit will be 0.

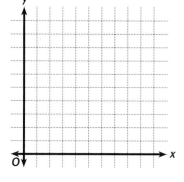

5. Work together to find a pricing strategy that will produce a larger profit for more vases, and write the revenue function.

34 Cooperative-Learning Activity   Algebra 2

NAME _____ CLASS _____ DATE _____

Cooperative-Learning Activity
6.1 Comparing Population Rates

Group members: 2

Materials: calculator

Responsibilities: Your group will analyze population growth for four United States continental regions and predict their populations in the years 2000 and 2015. You will also guess which regions will overtake others in population in the years 2000 and 2010.

Preparation: The table gives the populations of four regions of the United States in 1980 and the growth rate from 1980 to 1990.

	1980	Growth Rate	1990
Northeast	49,135,000	3.41%	
Midwest	58,866,000	1.36%	
South	75,372,000	13.8%	
West	43,172,000	22.26%	

Procedures:

1. One member estimates the population, to the nearest thousand, in 1990 for the Northeast and the Midwest. The other member does the same for the other two regions. Record this information in the table above.

2. Examine the completed table. Discuss any population shifts that you observe. Write a summary of the group's discussion on another sheet of paper.

3. Assuming the growth rates stay the same, predict any changes in the populations in 2000 and in 2010. Guess what the rankings by population will be in 2000 and in 2010, and write your rankings on another sheet of paper.

4. Complete the table below.

	year 2000	year 2010
Northeast		
Midwest		
South		
West		

5. How do your results compare with your predictions?

_____

_____

Algebra 2   **Cooperative-Learning Activity**   **35**

NAME _____ CLASS _____ DATE _____

Cooperative-Learning Activity
6.2 Water Mixtures and Exponential Functions

Group members: 4

Materials: tap water, ice water, hot water, effervescent tablets (dental or aspirin), Celsius thermometer, 5 clear plastic cups, stopwatch, measuring cup, pencil

Responsibilities: Conduct an experiment to model exponential decay.

Preparation: Pour 5 ounces of ice water in one cup, 5 ounces of tap water in another cup, and 5 ounces of hot water in a third cup. In a fourth cup, mix 2.5 ounces of tap water with 2.5 ounces of ice water. In the fifth cup, mix 2.5 ounces of tap water with 2.5 ounces of hot water.

Procedures:
1. Measure and record the temperature in each cup in the table in Procedure 3.
2. One member acts as a timekeeper and instructs the other members to drop an effervescent tablet into each cup at the same time.
3. Record the time in seconds when each tablet stops fizzing.

Temperature in °C (x)	
Time in seconds (y)	

4. Graph the data on the given coordinate axes.

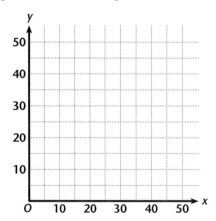

5. Use the graph to extend the pattern. Predict the amount of time for a temperature that is halfway between that of the hot water and that of the tap water.

_____

6. Compare the graph with the graph of $y = 2^x$. How are the graphs alike? How are they different?

_____

7. Discuss the relationship between temperature and reaction time, and write a description of the relationship.

_____

NAME _____ CLASS _____ DATE _____

Cooperative-Learning Activity
6.3 Logarithm Graph

Group members: 2

Materials: tracing paper, transparent tape, scissors

Roles: Exponential Graph Maker   draws graph of 2^x by plotting points
Logarithmic Graph Maker   reflects graph of 2^x across the line $y = x$

Preparation: Your group will graph an exponential function and use that function to create a logarithmic function. The Graph Makers will find logs of base 2 using the graph.

Procedures:
1. Each member chooses a role.

2. Given $y = 2^x$, complete the table below.

x	−4	−3	−2	−1	0	1	2	3
y								

3. The Exponential Graph Maker plots the points on the grid and draws a smooth curve through the points.

4. The Logarithmic Graph Maker plots the reflection of each point across the line $y = x$. Then he or she draws a smooth curve through those points, and writes an equation to describe this graph.

_____

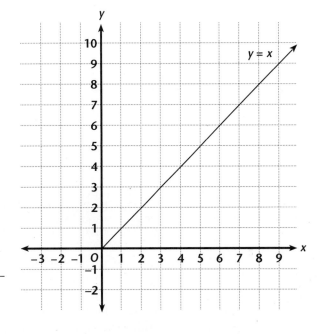

5. Discuss the relationship between the two graphs and the two related functions.

_____

_____

_____

NAME _____ CLASS _____ DATE _____

Cooperative-Learning Activity
6.4 Logarithm Relays

Group members: 6

Materials: index cards

Roles: **Logarithm Expert**   rewrites expressions containing logarithms as a single logarithm

Exponent Expert   rewrites equations with logarithms as equations with exponents

Solver   solves equations

Preparation: Teams will participate in a relay race. The relay will involve the following five equations:

A. $\log_{10} 20 + \log_{10} 5 = x$

B. $\log_2 12 - \log_2 3 = x$

C. $\log_3 3 + \log_3 x = 1$

D. $\log_8 20 - \log_8 40 = x$

E. $\log_3 27 + \log_3 3 - \log_3 x = 1$

Procedures:
1. The group divides into 2 teams. Each team has a Logarithm Expert, an Exponent Expert, and a Solver.

2. The Logarithm Expert writes equation A on an index card.

   a. The Logarithm Expert rewrites the expression as an expression involving a single logarithm. Then he or she passes the card to the Exponent Expert.

   b. The Exponent Expert rewrites the equation as an exponential equation. The the Exponent Expert passes the card to the Solver.

   c. The Solver solves the equation for x.

3. As soon as the Logarithm Expert completes work on equation A, he or she moves on to Equation B and Procedure 2 is repeated.

4. When your team has finished all five problems, check your final answers. Your team can win only if you get all 5 problems correct. If you find an incorrect answer, work together to find the correct answer.

5. The two Solvers collaborate to check all the work of both teams. Points are awarded as follows.

   Logarithmic equation correct: 3 points

   Exponential equation correct: 4 points

   Value of x correct: 3 points

   The team with the most points wins.

NAME _____ CLASS _____ DATE _____

Cooperative-Learning Activity
6.5 Investigating Sound Intensities

Group members: 2

Materials: no special materials

Responsibilities: Think about the following question: If the relative intensity of one sound is twice that of another sound, does that mean it is twice as loud?

Preparation: The table shows the decibel readings for various sounds.

Sound	Decibels (R)
Whisper	20
Bedroom	30
City living room	40
Friendly conversation	60
Hair dryer	80

Recall that $R = 10 \log \left(\dfrac{I}{I_0}\right)$ gives the relative intensity, R, in decibels of a sound whose intensity is I. The quantity I_0 is the intensity of the threshold of hearing.

Procedures:

1. Discuss and answer the question above. You must discuss what it means to be "twice as loud."

   _____

2. Each member finds two sounds such that the relative intensity of one is twice the relative intensity of the other. Each member then uses the equation to find how each of the sound intensities relates to the threshold of hearing, I_0.

Sound	Sound Intensity
Whisper	
Bedroom	
City living room	
Friendly conversation	
Hair dryer	

3. Share your results and discuss how your findings support or contradict your answer in Procedure 1.

   _____

   _____

Algebra 2   Cooperative-Learning Activity

NAME _____ CLASS _____ DATE _____

Cooperative-Learning Activity
6.6 Becoming a Millionaire

Group members: 2

Materials: calculator

Responsibilities: Your group will determine an investment strategy to acquire a million dollars in your lifetime starting from a given initial investment.

Preparation: Assume that you have $5000 to invest in an investment that is compounded continuously.

Procedures:
1. Discuss how large an interest rate would be needed to let an account grow to $1,000,000. One member writes down the guess.

   _____

2. Write an equation to determine how much time it will take for an account to amount to $1,000,000.

   _____

3. Each member substitutes a different interest rate in the equation and solves for t. Write your equations in the space below, and compare.

4. Each member chooses a larger interest rate, and Procedures 1–3 are repeated. Use the space below for your work.

5. Discuss interest rates at banks and how realistic the values found in Procedures 3 and 4 are. Write a summary of the discussion.

   _____
   _____

40 Cooperative-Learning Activity Algebra 2

NAME _____ CLASS _____ DATE _____

Cooperative-Learning Activity
6.7 Sorting Out Equations

Group members: 2

Materials: no special materials

Responsibilities: You will explore equations of the form $d = ab^c$. In each equation, three of the four quantities are fixed numbers and the fourth one is a variable.

Preparation: Use the four equations below.

  A. $d = 2(3^2)$ B. $12 = a(3^2)$

  C. $25 = 3(b^3)$ D. $130 = 5(3^c)$

Procedures:

1. One member of the group is assigned equations A and D.
   The second member of the group is assigned equations B and C.

2. The first member identifies the type of equation encountered and then writes the solution.

   $d = 2(3^2)$: Classification _____ Solution _____

   $130 = 5(3^c)$: Classification _____ Solution _____

3. The second member identifies the type of equation encountered and then writes its solution.

   $12 = a(3^2)$: Classification _____ Solution _____

   $25 = 3(b^3)$: Classification _____ Solution _____

4. When are logarithms needed to solve an equation of the form $d = ab^c$, when three of the four quantities a, b, c, and d are fixed numbers and one is a variable?

   _____

   _____

5. Consider an investment problem involving an initial deposit, a rate of interest with compounding, a final accumulation amount, and a period of time in which the initial deposit becomes the final amount. Under what conditions will a problem involving these quantities require logarithms to find the solution?

   _____

Algebra 2   Cooperative-Learning Activity   41

NAME _____ CLASS _____ DATE _____

Cooperative-Learning Activity
7.1 Comparing Annuities

Group members: 2

Materials: calculator

Responsibilities: You will compare the values of two different annuities.

Preparation: An insurance company offers two types of annuities, each of which requires an investment of $5000 over time.

- The first type invests $500 at the beginning of each year for 10 years. The account pays 4% interest compounded annually.

- The second type invests $1000 at the beginning of each year for 5 years. This account also pays 4% interest compounded annually.

Your group will find the final value of each account to see if there is a difference.

Procedures:

1. Discuss which annuity you think will be worth more at the time of the last payment.

   _____

   _____

2. Each member chooses a different plan and writes both a polynomial in x and a sum involving powers of 1.04 to model the plan.

   Type 1 _____

   _____

   Type 2 _____

   _____

3. Each member evaluates, to the nearest dollar, the polynomial written by the other member.

   Type 1 _____ Type 2 _____

4. Discuss the results and how they compare with the guess made in Procedure 1. What are the advantages and disadvantages of each type?

   _____

5. How could the insurance company change the annuities so that the advantages and disadvantages are more obvious?

   _____

   _____

42   Cooperative-Learning Activity   Algebra 2

NAME _____ CLASS _____ DATE _____

Cooperative-Learning Activity
7.2 Comparing Credit and Unemployment

Group members: 3

Materials: graphics calculators

Roles: **Credit Analyst** finds a quartic regression equation and graph for credit collection data

Unemployment Analyst finds a quartic regression equation and graph for unemployment data

Auditor checks equations of both analysts

Preparation: The table below gives data for past due credit payments and unemployment in the U.S. from 1990 to 1996. Credit amounts are given in billions of dollars. Numbers of unemployed persons are given in thousands.

Year	1990	1991	1992	1993	1994	1995	1996
Credit amount	751.9	745.0	756.9	807.1	925.0	1131.9	1224.4
People unemployed	7627	9375	10,491	9747	8745	8137	8024

Procedures:
1. The Credit Analyst and Unemployment Analyst find and graph quartic regression equations using $x = 1$ for 1990, $x = 2$ for 1991, and so on. While they are finding the graphs, the Auditor checks their equations.

   Credit: _____

   Unemployment: _____

2. The Analysts sketch the graphs on the grids provided. To approximate coordinates of points on the graphics calculator display, use TRACE.

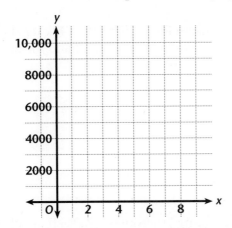
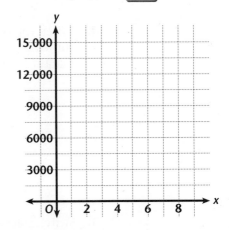

3. Look for any relationship between the outstanding credit and the unemployment from 1990 to 1996. The Auditor summarizes the discussion.

_____

Algebra 2   Cooperative-Learning Activity   **43**

NAME _____ CLASS _____ DATE _____

Cooperative-Learning Activity
7.3 Synthetic Division

Group members: 3

Materials: number cube, coin

Roles: Director determines a divisor

Worker follows the directions to complete synthetic division

Checker rewrites a problem as the product of two polynomials plus a remainder, and multiplies and adds to check the answer

Preparation: You will divide $P(x) = 2x^4 + 3x^3 - 2x^2 + 7x + 6$ by a factor determined by the Director.

Procedures:
1. Each member chooses a role.

2. The Director rolls the number cube and tosses the coin to determine a in the divisor $x - a$. The value of a is positive if the coin shows heads and negative if the coin shows tails. The numerical value of a is the number showing on the number cube.

3. The Director gives the Worker step-by-step instructions to divide $P(x)$ by $x - a$ by using synthetic division. The Checker checks each calculation as it is made. The Worker shows the work below.

4. When the synthetic division is complete, the Checker checks the division by showing that $P(x)$ is a product of the quotient and the divisor, plus the remainder. The Checker shows the work below.

5. Switch roles twice, repeating Procedures 1–4. Each member should assume each role once.

44 Cooperative-Learning Activity Algebra 2

NAME _____ CLASS _____ DATE _____

Cooperative-Learning Activity
7.4 Investigating Scuba Tanks

Group members: 2

Materials: measuring tape

Roles: Measurer measures another group member from the waist to the neck

Calculator finds the radius of a scuba tank for different volumes and heights

Preparation: The tanks used in scuba diving are in the shape of a cylinder with a hemisphere on the top. Tanks are available in different volumes. The total volume can be represented by the equation below.

$$V = \frac{2}{3}\pi r^3 + \pi r^2 h$$

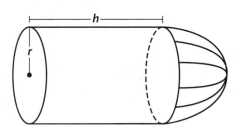

Procedures:
1. Each member chooses a role.

2. The Measurer finds a value for the height of a scuba tank by measuring another group member from the waist to the neck. The Measurer gives the height in inches. _____

3. The Calculator uses the height in Procedure 2 to find the radius of a scuba tank that has a volume of 2000 cubic inches. The Calculator gives the radius to the nearest tenth of an inch. _____

4. Switch roles and repeat Procedures 2 and 3 for a 2500-cubic-inch tank.

   height: _____ radius: _____

5. Discuss whether it is more practical to change the radius or the height of the scuba tank to obtain different volumes. A pressurized tank similar to the one in Procedure 3 holds 64,000 cubic inches of pressurized air. Calculate how much more pressurized air is contained in this scuba tank than in the non-pressurized tank.

   _____

   _____

   _____

   _____

Algebra 2   **Cooperative-Learning Activity** **45**

NAME _____ CLASS _____ DATE _____

Cooperative-Learning Activity

7.5 Finding Rational Zeros of Polynomial Functions

Group members: 3

Materials: graphics calculators

Roles: Factor Finder uses the Rational Root Theorem to find the possible roots of a given polynomial

Grapher graphs the function to approximate the roots

Evaluator evaluates the function at the points agreed upon by the group as possible roots

Preparation: Your group will work together to find the rational roots of the polynomial equation below.

$$y = -8x^3 - 26x^2 - 3x + 9$$

Recall that the Rational Root Theorem states that if $\frac{p}{q}$ is a rational root of $P(x) = 0$, then p is a factor of the constant term of $P(x)$ and q is a factor of the leading coefficient of $P(x)$.

Procedures:

1. Each member chooses a role.

2. The Factor Finder lists all of the possible values of $\frac{p}{q}$ that could be rational roots of $y = -8x^3 - 26x^2 - 3x + 9$.

   _____

3. The Grapher graphs the related function on the graphics calculator, and then reproduces a careful sketch on the grid provided.

   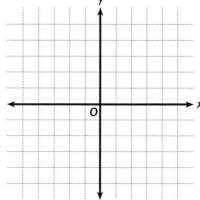

4. Together, look at the graph and the list of possible roots from Procedure 2, and guess the roots. The Evaluator substitutes the guesses into the polynomial to see if the value is 0. If not, look at the graph and the possible roots again, and make a new guess.

   _____

5. Discuss how you can use the graph and the process of elimination to rule out possibilities from the list made in Procedure 2.

   _____

   _____

46 Cooperative-Learning Activity   Algebra 2

NAME _____ CLASS _____ DATE _____

Cooperative-Learning Activity
8.1 Creating Scale Models

Group members: 4–6

Materials: calculator, ruler, paper, pencil

Responsibilities: Your group will determine the correct size for a model of the solar system.

Preparation: In a scale model, the ratio of the model dimensions to the real dimensions must be a constant. This constant is called the *scale factor*.

In this activity, your group will use a scale factor to calculate the correct sizes for models of the planets and the correctly scaled distances between the planets.

The table below approximates the planet sizes and distances from the Sun.

Planet	Real diameter (km)	Real distance from the Sun (km)
Sun	1,391,000	0.00
Mercury	4880	5.8×10^7
Venus	12,100	1.1×10^8
Earth	12,756	1.5×10^8
Mars	6791	2.3×10^8
Jupiter	143,200	7.8×10^8
Saturn	120,000	1.4×10^9
Uranus	51,800	2.9×10^9
Neptune	49,500	4.5×10^9
Pluto	3000	5.9×10^9

Procedures:
1. Create a table like the one above with real diameters of the planets in one row and real distances from the Sun in another row. Then create two new rows, one for model diameter in centimeters and one for model distance from the Sun in meters.

2. Use the scale factor 1.8×10^{-5} centimeters per kilometer for the model diameter and 1.8×10^{-7} meters per kilometer for model distance. Note that these are the same scale factors. As a group, calculate the dimensions for your scale model and record the results in your table.

3. Imagine modeling the Sun with a basketball. Name an object that could model Jupiter. _____

4. If the model of the Sun were placed on your desk, where would the model of Pluto be located? _____

5. If the Sun were modeled by a basketball, what could you use to model Earth? If the basketball were placed on your desk, where would Earth be located?

Algebra 2   Cooperative-Learning Activity   **47**

NAME _____ CLASS _____ DATE _____

Cooperative-Learning Activity
8.2 Transformations of the Reciprocal Function

Group members: 3

Roles: Divider   rewrites a rational function using long division or synthetic division
Transformer   identifies transformations made to the reciprocal function
Grapher   sketches a graph using transformations

Preparation: Each of the following rational functions can be rewritten as a transformation of the reciprocal function, $y = \frac{1}{x}$.

A. $y = \frac{2x + 4}{x + 4}$ B. $y = \frac{-2x + 9}{x + 3}$ C. $y = \frac{5x^2 - x}{x^2 - x}$

Procedures:
1. The Divider uses long division or synthetic division to rewrite function A.

2. The Transformer lists the transformations of the reciprocal function that produce function A.

3. The Grapher sketches the graph of function A.

4. Switch roles and repeat Procedures 2–4 for function B.

5. Switch roles again and repeat Procedures 2–4 for function C.

6. Can all rational functions be rewritten as transformations of the reciprocal function? If so, explain why. If not, describe which rational functions can be written in this way.

A.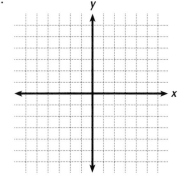

Function _____

Transformations

_____

_____

B.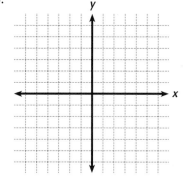

Function _____

Transformations

_____

_____

C.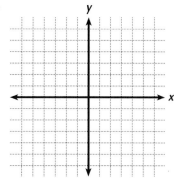

Function _____

Transformations

_____

_____

NAME _____ CLASS _____ DATE _____

Cooperative-Learning Activity
8.3 Comparing Areas of Rectangles

Group members: 3

Roles: **Rectangle Maker** chooses two binomials in x to model the lengths of the sides of a rectangle

Ratio Maker chooses which rectangle will have the larger area and writes a ratio of the areas

Solver simplifies and solves the inequality

Preparation: Your group will investigate comparisons of rectangle areas for various values of x by using a ratio to compare the areas. The inequality below will be used in the procedures that follow.

If the area of rectangle A is more than r times the area of rectangle B, then $\dfrac{\text{length of rectangle A} \times \text{width of rectangle A}}{\text{length of rectangle B} \times \text{width of rectangle B}} > r.$

Procedures:
1. Each member chooses a role.

2. The Rectangle Maker chooses two binomials to represent the sides of rectangle A and another binomial to represent the length of rectangle B. The width of rectangle B should be the same as one of the sides of rectangle A. The Rectangle Maker then writes an expression for the area of each rectangle.

   rectangle A _____   rectangle B _____

3. The Ratio Maker chooses which rectangle has a larger area. He or she also chooses a comparison factor. For example, the Ratio Maker might decide that rectangle A is at least twice as large as rectangle B. The Ratio Maker then writes an inequality to represent the situation.

   _____

4. The Solver simplifies the rational expression from Procedure 3 and finds values of x for which the inequality is true. If there are no such values, then the Solver reverses the direction of the inequality and solves the new inequality.

   _____

5. Rotate roles and repeat Procedures 1–4 two more times.

   rectangle A _____   rectangle B _____

   rectangle A _____   rectangle B _____

Algebra 2   Cooperative-Learning Activity   **49**

NAME _____ CLASS _____ DATE _____

Cooperative-Learning Activity
8.4 Cruising Up the River

Group members: 2

Materials: no special materials

Responsibilities: You will find an expression for the time it takes a motorboat to complete a trip from a campsite to a swimming beach located 2 nautical miles up the river.

Preparation: The speed of the motorboat is 8 nautical miles per hour. The time it takes for the boat to go upstream will be different than the time it takes to go downstream.

Recall that distance = rate × time and that time = $\frac{\text{distance}}{\text{rate}}$.

Procedures:

1. Discuss what is known and what is unknown in this situation.

   known: _____

   unknown: _____

2. Use the variable c to represent the speed of the current. One member represents the time of the trip upstream in terms of the rate. The other member represents the time of the trip downstream.

   upstream time expression: _____

   downstream time expression: _____

3. Discuss how to find an expression for the total time to the beach and back, ignoring any time spent at the beach. One member writes down the expression.

   total trip time: _____

4. **a.** Find the time for currents whose speeds are 1, 2, and 3 nautical miles per hour.

   1 nautical mile per hour: _____

   2 nautical miles per hour: _____

   3 nautical miles per hour: _____

   b. Discuss what the maximum value of the current can be for the motorboat to navigate on the river. What happens if the current is greater than this rate?

   _____

50 Cooperative-Learning Activity   Algebra 2

NAME _____ CLASS _____ DATE _____

Cooperative-Learning Activity
8.5 Analyzing a Trip

Group members: 3

Materials: graphics calculator

Roles: "Father" fills in table for father and writes equation

"Mother" fills in table for mother and simplifies equation

"Son" fills in table for son and checks simplification

Preparation: A family plans to drive 2000 miles in 33 hours. The father will drive half of the distance, and the mother and son will share equally the remaining distance.

In general, the father drives the fastest, the mother drives 10 miles per hour slower than the father, and the son's rate is halfway between the mother's and father's rates.

You will find the rates for each of the drivers.

Procedures:
1. Each member chooses a role.

2. Discuss how to find the distance for each person, and complete that column in the table.

	r	t	d
Father			
Mother			
Son			
Total		33	2000

3. Express each rate in terms of the father's rate, which is represented by x, and record the expressions in the table.

4. Discuss how to complete the time column.

5. The Father writes an equation for the total time. _____

6. The Mother simplifies the equation, and the Son checks the simplification.

_____

7. Discuss how a graphics calculator can be used to solve the equation, including a discussion of a reasonable answer, and solve the equation.

_____

Algebra 2   **Cooperative-Learning Activity**   **51**

NAME _____ CLASS _____ DATE _____

Cooperative-Learning Activity
8.6 Building a Transformation

Group members: 4

Materials: pencil, index cards, number cube, coin, grid paper

Roles: **Roller** rolls number cube and tosses coin to determine specific translations and stretches

Dealer shuffles cards and places them face down

Transformer writes equation of the transformation

Grapher graphs the transformation

Preparation: Prepare the index cards by writing each transformation on a card:

reflection   horizontal translation   vertical translation   vertical stretch

You will be building transformations of $y = \sqrt{x}$.

Procedures:
1. Each member chooses a role. The Dealer shuffles the cards and places them face down.

2. The Dealer picks two cards, and the Roller rolls the number cube three times for each transformation to determine a, b, and c in $y = a\sqrt{x - b} + c$.
The Dealer tosses the coin to find the sign of a, b, and c. Heads is positive and tails is negative.

3. The Transformer writes a function that satisfies the given transformations. The Grapher graphs the transformed function.

   _____

4. Repeat Procedures 2 and 3, switching roles until each member has sketched a graph. Write the functions in the space below.

   _____

   _____

   _____

Cooperative-Learning Activity
8.7 Spinning Radicals

Group members: 3

Materials: paper clips, pencil, number cube

Roles: **Leader** rolls number cube to determine operation and checks work of Players

Players spin the spinners to find two expressions that each represent square roots or cube roots

Preparation: To use the spinners, hold a pencil through one end of a paper clip and place the pencil point at the center of the spinner. Then spin the paper clip around the pencil. Spin again if the paper clip lands on a line.
The Players will perform operations on two expressions determined by the Leader.

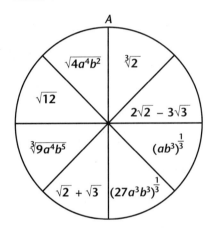

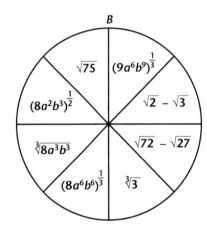

Procedures:
1. Each member chooses a role. The Leader rolls the number cube until a 1, 2, 3, or 4 is obtained and calls out the operation as follows:
   1: addition   2: subtraction   3: multiplication   4: division

2. The first Player spins Spinner A and records the expression in the table. Then he or she spins Spinner B until an expression with the same root is determined. The second Player then spins in the same manner.

Player	Operation	Spinner A	Spinner B	Answer	Point

3. Each Player performs the operation indicated in his or her table and writes the answer in simplified form. In some cases, it may not be possible to simplify beyond writing the indicated problem.

4. The Leader awards 2 points for a correct answer, 1 point for a correct answer that can be simplified further and 0 points for an incorrect answer.

5. Switch roles and play until one member has a total of 10 points.

Algebra 2   Cooperative-Learning Activity   53

NAME _____ CLASS _____ DATE _____

Cooperative-Learning Activity
8.8 Revolving Planets

Group members: 3

Materials: calculators

Roles: **Statistician** finds a value for C using data about Earth

Mathematician solves the equation for r

Engineer finds the average distance between Venus and the Sun

Preparation: Kepler's Third Law of Planetary Motion states that "the ratio of the average distance from the Sun cubed to the period squared is the same constant for all planets." This can be written as stated below.

$$\frac{(r_p)^3}{(T_p)^2} = C,$$ where r_p represents the radius of the planet's orbit and T_p represents the time it takes the planet to make one full revolution.

For Earth, one full revolution takes 365.256 days. The average distance between Earth and the Sun is 149.6×10^6 kilometers.

Procedures:

1. Discuss what type of variation is represented in Kepler's Law equation.

   _____

   _____

2. The Statistician finds the value of C for Earth while the Mathematician solves Kepler's Law equation for T_p.

   _____

   _____

3. The Statistician and Mathematician check each other's work.

4. The distance between Venus and the Sun is 108.2×10^6 kilometers. Let T_V represent the time it takes for Venus to make one complete revolution around the Sun. The Engineer solves the equation for T_V.

   _____

5. Discuss what data you would need to calculate the periods of the other planets' orbits around the Sun.

   _____

   _____

   _____

NAME _____ CLASS _____ DATE _____

Cooperative-Learning Activity

9.1 Meet You Halfway There

Group members: 2

Materials: paper clip

Responsibilities: You will determine the location of a meeting place that is located at the midpoint of two locations.

Preparation: A coordinate system has been used as a map for Wonderful Town. Several buildings and a park have been identified on the grid. Each site has integer coordinates.

Use a paper clip as a pointer by placing a pencil point through the paper clip at the center of the spinner.

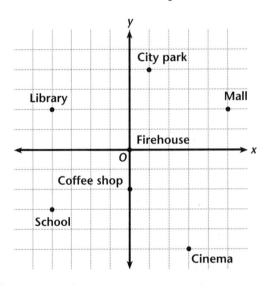
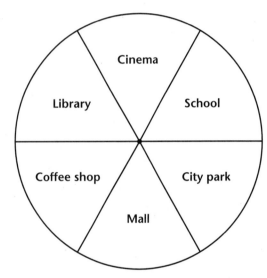

Procedures:

1. Each member spins the paper clip to determine his or her location in Wonderful Town. If both locations are the same, one member spins again. Record each member's location in the table below.

2. The members are to meet halfway between the two sites. Each member finds the coordinates of the meeting point.

3. Repeat Procedures 1 and 2 four times and enter the information in the table below.

	Coordinates
Member 1	
Member 2	
Meeting point	

	Coordinates			
Member 1				
Member 2				
Meeting point				

Algebra 2   Cooperative-Learning Activity   **55**

NAME _____ CLASS _____ DATE _____

Cooperative-Learning Activity
9.2 Studying the Flashlight

Group members: 2

Materials: no special materials

Roles: **Optical Analyzer** analyzes a drawing of a flashlight

Modeler writes an equation for a parabola

Preparation: You will investigate the flashlight and model different-shaped flashlights.

Procedures:
1. Each member chooses a role.

2. The Optical Analyzer studies the diagram below. Then he or she identifies the type of curve, and the meanings of point O, point F, and the x-axis in this context.

   curve: _____

   points O and F: _____

   x-axis: _____

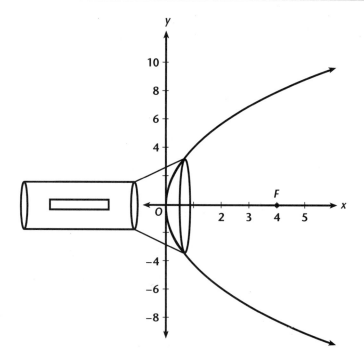

3. The Modeler writes an equation for the curve. _____

4. Discuss how to change the shape of the curve by moving F.

_____

_____

56 Cooperative-Learning Activity   Algebra 2

NAME _____ CLASS _____ DATE _____

Cooperative-Learning Activity
9.3 Intersecting Circles

Group members: 3

Materials: no special materials

Responsibilities: You will graph circles by completing the square to find their radii and centers.

Preparation: One of the following sets of circles contains three circles that do not intersect. You will identify that set.

Set A: $\begin{cases} x^2 - 6x + y^2 - 8y + 16 = 0 \\ x^2 + 2x + y^2 - 3y - 5.75 = 0 \\ x^2 - 6x + y^2 + 2y + 1 = 0 \end{cases}$

Set B: $\begin{cases} x^2 - 10x + y^2 - y + 21.25 = 0 \\ x^2 + 2x + y^2 + 4y + 1 = 0 \\ x^2 - 2x + y^2 - 6y + 6 = 0 \end{cases}$

Set C: $\begin{cases} x^2 + y^2 - 2y = 0 \\ x^2 + 4x + y^2 + 2y + 4 = 0 \\ x^2 - 2x + y^2 - 2y + 1 = 0 \end{cases}$

Procedures:

1. Examine the equations that make up each set.
   Guess which set of equations has no intersecting circles.

   _____

2. For the set of equations that you chose in Procedure 1, each member completes the square for one equation and graphs that equation on grid A at left below.

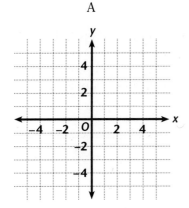
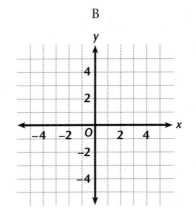
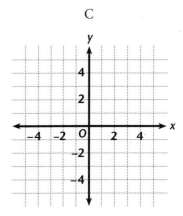

3. If the diagram at left above shows intersecting circles, repeat Procedure 1 for the remaining sets of equations. Then repeat Procedure 2 for the set of equations that you believe has nonintersecting circles.

4. Identify the set of equations that contains nonintersecting circles and sketch those circles on the grid at right above. _____

Algebra 2   **Cooperative-Learning Activity**

NAME _____ CLASS _____ DATE _____

Cooperative-Learning Activity
9.4 Analyzing an Ellipse

Group members: 4

Materials: no special materials

Roles: **Leader** writes the equation of an ellipse in standard form and checks all work done by other members

Point Plotter identifies and plots vertices, foci, and center of ellipse

Analyzer identifies and sketches the major and minor axes

Sketcher sketches the ellipse

Preparation: Your group will work together to identify the foci, vertices, and axes of ellipses. You will then sketch the graph.

Procedures:
1. Each member chooses a role.

2. The Leader writes the equation of an ellipse in standard form. (The ellipse should fit on the grid at left below.) _____

3. The Analyzer finds the major and minor axes and sketches them on the grid at left below.

   _____

4. The Point Plotter finds the center, foci, vertices, and covertices of the ellipse and plots them on the grid at left below.

   _____

5. The Sketcher sketches the graph of the ellipse on the grid at left below.

6. Switch roles twice, using the grids below. Graph at least one ellipse with a center other than the origin and at least one ellipse with its major axis parallel to the y-axis.

   Equation: _____   Equation: _____

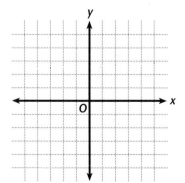 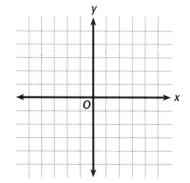 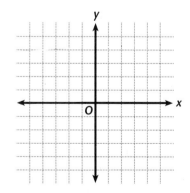

58   Cooperative-Learning Activity   Algebra 2

NAME _____ CLASS _____ DATE _____

Cooperative-Learning Activity
9.5 Comparing Hyperbolas

Group members: 2

Materials: differently colored pencils

Responsibilities: You will graph pairs of hyperbolas to examine any relationships between the graphs in each pair.

Preparation: The following are equations of hyperbolas in which x and y have been exchanged in the second equation.

Set A. $\dfrac{x^2}{16} - \dfrac{y^2}{25} = 1 \qquad \dfrac{y^2}{16} - \dfrac{x^2}{25} = 1$

Set B. $\dfrac{x^2}{9} - \dfrac{(y+2)^2}{4} = 1 \qquad \dfrac{y^2}{9} - \dfrac{(x+2)^2}{4} = 1$

Set C. $\dfrac{(x-5)^2}{4} - \dfrac{(y+1)^2}{16} = 1 \qquad \dfrac{(y-5)^2}{4} - \dfrac{(x+1)^2}{16} = 1$

Procedures:

1. For Set A, each member uses a different color to graph one equation on the grid at left below.

2. Discuss how the two graphs are related and write a summary of the relationship.

    Set A: _____

3. Repeat Procedures 1 and 2 for Sets B and C.

    Set B: _____

    Set C: _____

4. Make a generalization about the effect of interchanging x and y in the equation of a hyperbola.

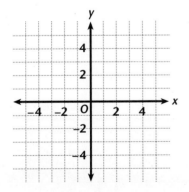
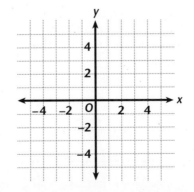
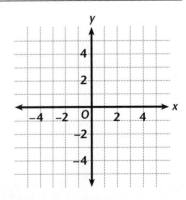

Algebra 2 — Cooperative-Learning Activity

NAME _____ CLASS _____ DATE _____

Cooperative-Learning Activity
9.6 Thunder and Lightning

Group members: 3

Materials: graphics calculator

Roles: **Modeler** writes an equation to describe the situation

Solver solve the equations using the graphics calculator

Checker checks the work and makes a drawing of the situation on the grid

Preparation: Two friends, Tosha and Emilio, are talking on the phone when they both notice a flash of lightning at the same time. Tosha hears the thunder 4 seconds later. Emilio hears the thunder 6 seconds later.

Using 0.33 kilometers per second as the speed of sound and the grid showing the locations of Tosha's and Emilio's houses, you will locate the possible locations of the lightning strike.

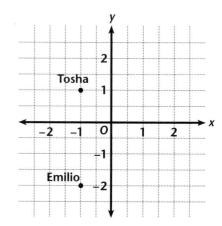

Procedures: 1. Each member chooses a role.

2. Discuss the problem by determining what is known and unknown and making any necessary calculations. Discuss the possible location of the lightning from the point of view of Tosha and of Emilio. The Modeler summarizes the discussion by writing equations that model the situation.

_____

_____

_____

3. The Solver solves the equations. _____

4. The Checker makes a sketch of the situation on the grid, checks the calculations, and makes sure that the answer is reasonable.

_____

_____

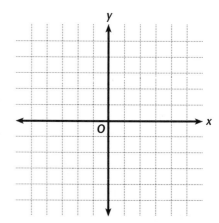

60 Cooperative-Learning Activity Algebra 2

Cooperative-Learning Activity

10.1 Coin-Tossing Experiment

Group members: 3

Materials: 3 different coins, such as dime, nickel, and penny

Roles: Coin Tosser   tosses 3 coins at one time

Recorder   records results

Calculator   calculates theoretical probability

Preparation: You will perform a probability experiment involving three coins to find the experimental probability of tossing a certain number of heads.

Procedures:

1. The Calculator writes all the possible outcomes of the experiment in the table at right in order to determine the size of the sample space. Then he or she calculates the theoretical probability of tossing three heads.

0 heads	
1 head	
2 heads	
3 heads	

2. The Coin Tosser tosses the three coins. The Recorder keeps track of the number of tosses and the outcomes. The Coin Tosser repeats the tosses 20 times.

Trial	Outcome	Trial	Outcome
1		11	
2		12	
3		13	
4		14	
5		15	
6		16	
7		17	
8		18	
9		19	
10		20	

3. The Recorder calculates the experimental probability, and all group members compare the theoretical and experimental probabilities. Discuss what should happen as the number of trials increases.

	Probability
0 heads	
1 head	
2 heads	
3 heads	

Algebra 2 — Cooperative-Learning Activity

NAME _____ CLASS _____ DATE _____

Cooperative-Learning Activity
10.2 Line-Up

Group members: 5

Materials: no special materials

Roles: **Photographer** tells models where to stand

Models follow Photographer's instructions

Preparation: Your group is going to have a picture taken. You will find the number of possible arrangements of the people in the photograph. The Photographer will not be in the photograph at first.

Procedures:
1. The Photographer arranges the Models, and writes the arrangement below. Then he or she rearranges the Models until all possible orders have been found. He or she writes the total number of arrangements. _____

2. Discuss various other ways to achieve a complete listing of the arrangements of the four Models.

   _____
   _____
   _____

3. Consider including the Photographer in the picture. In each existing arrangement, how many places are there for him or her to stand? How many total arrangements are there including the Photographer?

   _____
   _____

4. Choose a new Photographer. He or she directs the four Models to stand around a circle. How many different arrangements are there?

   _____
   _____

NAME _____ CLASS _____ DATE _____

Cooperative-Learning Activity
10.3 Finding Triangles

Group members: 2

Materials: no special materials

Responsibilities: You will determine the number of triangles that contain a set of points on the circle.

Preparation: You will use the circle shown at right below.

Recall that △ABC can also be designated as follows:

△ACB △BAC △BCA △CAB △CBA

Therefore, △ABC, △ACB, △BAC, △BCA, △CAB, and △CBA are all considered to be the same triangle.

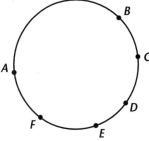

Procedures: 1. Determine how many triangles can be drawn using the six points shown here.

_____

2. Add a point G to the circle and repeat Procedure 1.

_____

3. Add a point H to the circle and repeat Procedure 1.

_____

4. Calculate the number of triangles that can be drawn using n points positioned around the circle and explain how you arrived at your answer.

_____
_____
_____
_____

5. Discuss how the number of triangles is related to permutations or combinations.

_____
_____

Algebra 2   Cooperative-Learning Activity   **63**

NAME _____ CLASS _____ DATE _____

Cooperative-Learning Activity
10.4 Favorite Music Survey

Group members: 4

Materials: no special materials

Roles: **Data Collector** compiles the results of a survey

Classifier decides whether an event is inclusive or mutually exclusive

Probability Expert finds probability of given events

Checker checks the work of the Classifier and Probability Expert

Preparation: Each member fills out the following survey.

Male ☐   Female ☐

What is your favorite kind of music? You may choose only one answer:

Rock and roll ☐   Rap ☐

Country western ☐   No favorite ☐

Procedures:
1. Each member chooses a role. The Data Collector completes the table.

	Male	Female	Total
Rock and roll			
Country western			
Rap			
No favorite			

2. For each of the following events, the Classifier determines whether the events are inclusive or mutually exclusive. The Probability Expert calculates the probability of the event, and the Checker checks the work.

   a. A member chosen at random is male or likes country western music.

   Classification _____   Probability _____

   b. A member chosen at random is female or likes rap music.

   Classification _____   Probability _____

3. Rotate roles so that there is a new Classifier, Probability Expert, and Checker. Repeat Procedure 2 with the following event.

   A member chosen at random likes country western music or has no favorite.

   Classification _____   Probability _____

NAME _____ CLASS _____ DATE _____

Cooperative-Learning Activity
10.5 Matching the Goal

Group members: 4

Materials: two number cubes, 11 one-inch square pieces of paper, a bag

Roles: **Probability Expert** draws goal and determines theoretical probability
Recorder records outcomes
Players roll number cubes

Preparation: Write each of the numbers from 2 to 12 inclusive on pieces of paper and put the papers in a bag. You will play a game with number cubes in which both players try to get the same sum as the goal, which is determined by the number drawn from the bag.

Let $P(A)$ represent the probability that Player 1 matches the goal.
Let $P(B)$ represent the probability that Player 2 matches the goal.

Procedures:
1. Each member chooses a role.

2. The Probability Expert draws a slip of paper from the bag to set the goal. The Probability Expert determines $P(A)$ and $P(B)$. Then he or she calculates $P(A \text{ and } B)$, the probability that both players will achieve the goal on their roll of the number cubes.

   _____

   _____

3. Each player rolls the number cubes. The Recorder records each sum in the table. The Players roll the number cubes several more times.

Goal:	Player 1	Player 2

4. Record the number of successful outcomes and compare the results with $P(A \text{ and } B)$ determined by the Probability Expert. In 10 trials, how many successful outcomes would be expected?

5. Switch roles until each member has been Probability Expert once and Recorder once.

6. Discuss how the goal affects $P(A \text{ and } B)$. Which goal has the highest probability?

   _____

   _____

Algebra 2   **Cooperative-Learning Activity**

NAME _____ CLASS _____ DATE _____

Cooperative-Learning Activity
10.6 Conditional Probability

Group members: 3

Materials: calculators

Roles: **P(A) Finder** finds the probability of event A

P(A and B) Finder finds the probability of event A and B

P(B|A) Finder finds the conditional probability

Preparation: You will use the data below to find conditional probabilities for the following situations:
X. Given that a person works in agriculture, he or she works 40 hours per week.
Y. Given that a person works in a non-agriculture industry, he or she works 14 hours or less per week.
Z. Given that a person works 15 to 29 hours per week, he or she works in agriculture.

In the following procedures, event A refers to the event for which information is given, and event B refers to the event subsequent to event A.

Weekly Hours Worked 1994 (in thousands)			
	Total	Agriculture	Non-agriculture
1 to 4 hours	1271	84	1187
5 to 14 hours	4992	262	4730
15 to 29 hours	15,115	493	14,623
30 to 34 hours	9473	225	9248
35 to 39 hours	8684	168	8516
40 hours	40,587	624	39,963
41 hours and over	37,319	1352	35,966
Total	117,441	3208	114,233

Procedures: 1. For situation X, identify events A and B.

event A: _____

event B: _____

2. Each member finds the appropriate probability as a fraction. P(B|A) Finder uses the probabilities provided by P(A) Finder and P(A and B) Finder.

Situation X: P(A) _____ P(A and B) _____ P(B|A) _____

3. Switch roles twice and repeat Procedures 1 and 2 for situations Y and Z.

Situation Y: P(A) _____ P(A and B) _____ P(B|A) _____

Situation Z: P(A) _____ P(A and B) _____ P(B|A) _____

NAME _____ CLASS _____ DATE _____

Cooperative-Learning Activity
10.7 Simulating the Collection of Mugs

Group members: 2

Materials: number cube

Roles: **Roller** rolls the number cube

Recorder records the outcomes

Preparation: You will simulate the following situation:
A local hamburger chain is offering three special mugs with the purchase of a hamburger. How many hamburgers do you have to buy to get a complete set of mugs?
Set up the simulation as follows:
The roll of the number cube represents the acquisition of a mug.
1 or 2 represents mug A   3 or 4 represents mug B   5 or 6 represents mug C

Procedures: 1. Each member chooses a role.

2. The Roller rolls the number cube. The Recorder writes the mug corresponding to the number rolled in the table below. If, for example, on Trial 1, the Roller shows 1, 2, 4, 2, 3, and 6, then mugs, A, A, B, A, B, and C, respectively, are acquired, and it took six rolls to acquire them all.

Trial number	Trial	Number of rolls
1	AABABC	6

Perform 10 trials. Remember that a trial must continue until all 3 mugs have been acquired.

3. Switch roles and conduct 10 more trials.

4. Estimate the number of hamburgers that must be purchased in order to get a complete set of three mugs.

_____

Algebra 2   Cooperative-Learning Activity   **67**

NAME _____ CLASS _____ DATE _____

Cooperative-Learning Activity
11.1 Recursive Relays

Group members: 6

Materials: 4 index cards, calculator

Responsibilities: Your group members will compete in a relay race.

Preparation: Divide into two teams of three members each. Each team prepares an index card for each of the following sequences:

Sequence A $a_1 = 18$ and $a_n = 3a_{n-1} + 5$, where $n > 1$

Sequence B $a_1 = -8$ and $a_n = -4a_{n-1} + 50$, where $n > 1$

You will compete to see which team can be the first to correctly evaluate the first ten terms in sequence A and their sum, $\sum_{j=1}^{10} a_j$.

Procedures:
1. Each team chooses the order in which the members will work the problems.

2. The first player of each team takes the index card for sequence A. He or she calculates a_2, writes it on the card under a_1, and gives the card to the next team member.

3. The next team member calculates a_3, writes it under a_2, and hands it to the last team member, who then calculates a_4 and writes it down.

4. The last team member hands the card to the first member, and the relay continues until a_{10} is calculated.

5. The member who calculates a_{10} hands the card to the next member, who must then calculate the sum of the terms, $\sum_{j=1}^{10} a_j$.

6. The first team to get the right answer for $\sum_{j=1}^{10} a_j$ is the winner.

7. Form two different teams and repeat Procedures 2–7 with sequence B.

NAME _____ CLASS _____ DATE _____

Cooperative-Learning Activity

11.2 Exploring the Buy or Lease Options

Group members: 3

Materials: calculator, grid paper

Roles: Buyer calculates costs associated with buying a car

Leasee calculates costs associated with leasing a car

Grapher graphs sequences associated with buying and with leasing

Preparation: Dawn is considering two options for acquiring a car.
Option A: She can purchase a car for $45,000. The value of the car decreases by $4000 per year.
Option B: She can lease a car for $200 per month. In this case, there is no loss in value due to depreciation.

Procedures: 1. Each member chooses a role.

2. a. The Buyer completes the table below to find the value of the car after the given number of years.

Years	0	1	2	3	4	5	6
Value							

b. The Buyer represents the value of the car as a sequence and identifies the sequence type.

_____

c. The Grapher represents the value of the car as a graph using either the table from part **a** or the sequence from part **b**.

3. a. The Leasee represents the total cost of leasing over n months as a sequence and identifies the sequence type.

_____

b. The Grapher represents the cost of leasing the car over n months as a graph.

4. Discuss the two sequences for buying and leasing, and explain how you know that each sequence is an arithmetic sequence.

_____

_____

_____

Algebra 2   Cooperative-Learning Activity   **69**

NAME _____ CLASS _____ DATE _____

Cooperative-Learning Activity
11.3 Arithmetic Series

Group members: 3

Materials: slips of paper numbered from 6 to 20 inclusive, a bag

Roles: **Random Number Generator** selects one of the slips of paper from the bag

Term Finder finds the nth term of the specified sequence when n is the number generated randomly and finds a formula for the nth term of a series

Sum Finder finds the sum

Preparation: One member cuts fifteen 1-inch squares. Another member numbers the squares from 6 to 20, inclusive. You will find a formula for the following arithmetic sequences, and you will find the value of the sum of the first n terms, where n is the number generated by the Random Number Generator.
   A. $5, 2, -1, -4, -7, \ldots$
   B. $15, 24, 33, 42, 51, \ldots$
   C. $-50, -37, -24, -11, 2, \ldots$

Procedures:
1. Each member chooses a role. The Random Number Generator generates a number n by pulling a piece of paper from the bag.

2. The Term Finder finds t_n for sequence A. The Sum Finder uses t_n to find the sum of the first n terms of the series related to sequence A. The Random Number Generator checks the work.

   n _____ t_n _____ S_n _____

3. The Random Number Generator draws another value of n and Procedure 2 is repeated.

   n _____ t_n _____ S_n _____

4. Switch roles so that no member has the same role as before and repeat Procedures 2 and 3 for sequence B.

   Sequence B: n _____ t_n _____ S_n _____

   n _____ t_n _____ S_n _____

5. Switch roles so that no member has the same role as before and repeat Procedures 2 and 3 for sequence C.

   Sequence C: n _____ t_n _____ S_n _____

   n _____ t_n _____ S_n _____

NAME _____ CLASS _____ DATE _____

Cooperative-Learning Activity
11.4 Chasing Shrinking Squares

Group members: 3

Materials: ruler

Roles: **Drawer** draws a sequence of squares
Area Finder finds the area of each square
Mathematician finds the exact length of the side of each square

Preparation: You will generate a sequence in which a square is determined by the midpoints of the sides of a previously drawn square. The starting square is quadrilateral ABCD.

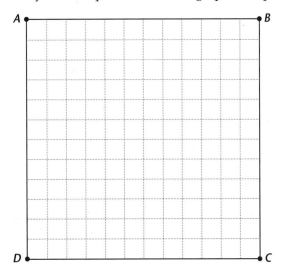

Procedures:

1. The Drawer finds the midpoint of each side of square ABCD by measuring the sides of the square. Then he or she creates a new square by drawing line segments connecting the midpoints.

2. The Mathematician finds the exact length of the sides of the new square and the Area Finder finds the area.

3. Repeat Procedures 2 and 3 for the next four squares in the sequence.

4. Complete the table below.

Exact length					
Area					

5. Determine whether the sequences are geometric, arithmetic, or neither. If a sequence is arithmetic, give d. If it is geometric, give r.

Lengths _____ d or r _____

Area _____ d or r _____

Algebra 2 Cooperative-Learning Activity **71**

NAME _____ CLASS _____ DATE _____

Cooperative-Learning Activity
11.5 Sequences and Series

Group Members: 3

Materials: calculators

Roles: Sequence Identifier — identifies whether a sequence is arithmetic, geometric, or neither

Arithmetic Expert — finds a specific term and sum of series for arithmetic sequences

Geometric Expert — finds a specific term and sum of series for geometric sequences

Preparation: You will use the following sequences:

Group A	Group B
1. 15, 30, 60, 120, …	1. 1, 4, 9, 25, …
2. 15, 16, 18, 21, 25, …	2. 100, 75, 50, 25, …
3. 15, 13, 11, 9, 7, …	3. 100, 75, 56.25, 42.1875, …
4. 8, −16, 32, −64, …	4. 0.1, −1, 10, −100, …
5. 8, −8, −24, −40, …	5. −0.1, 1, 2.1, 3.2, …

Procedures: 1. Each member chooses a role.

2. Examining the sequences in Group A, the Sequence Identifier sorts the sequences into arithmetic, geometric, or neither. Those that are arithmetic are given to the Arithmetic Expert, and those that are geometric are given to the Geometric Expert. Any sequences that are neither arithmetic nor geometric are discarded.

3. The Arithmetic Expert and Geometric Expert each find the 10th term and the sum of the first ten terms for the appropriate sequences.

            Arithmetic   Geometric

       t_{10} _____ S_{10} _____ t_{10} _____ S_{10} _____

4. Switch roles and repeat Procedures 2 and 3 for Group B.

            Arithmetic   Geometric

       t_{10} _____ S_{10} _____ t_{10} _____ S_{10} _____

NAME _____ CLASS _____ DATE _____

Cooperative-Learning Activity
11.6 Crossing the Finish Line

Group members: 3

Materials: no special materials

Roles: **Measurer** marks distances on the number line
Mover moves to points specified by the Measurer
Recorder makes a sketch showing the action of the Mover

Preparation: You will investigate how many steps it will require for the Mover to cross the finish line, given that each step the Mover takes is half the length of the step previously taken.

Procedures:
1. Each member chooses a role.

2. The Measurer finds the length of the number line below. The distance between two consecutive tick marks is one unit. The Mover points to the starting point, point A.

3. The Measurer marks the point halfway between the Mover and the finish line, point B. The Mover moves his or her pencil so that its point is on that point and the Recorder draws a line showing the distance that the Mover has moved, with a mark at the Mover's new position.

4. Procedure 3 is repeated as many more times as possible.

A ●————————————————————————————● B

5. Complete the table below.

Step number	1	2	3	4	5	n
Step length						

6. Discuss the statement below. Is this a true statement?

    The Mover will never reach point B.

    _____

    _____

7. Work together to write and evaluate an infinite geometric series to represent the total distance traveled by the Mover.

    _____

Algebra 2   Cooperative-Learning Activity   **73**

NAME _____ CLASS _____ DATE _____

Cooperative-Learning Activity
11.7 Pascal Diagonals

Group members: 3

Materials: no special materials

Roles: **Triangle Maker** extends Pascal's triangle and draws models of triangular numbers
Diagonal Selector draws a ring around part of a diagonal of the triangle
Sum Finder finds specified sums

Preparation: You will investigate patterns in the diagonals of Pascal's triangle.

```
                1
              1   1
            1   2   1
          1   3   3   1
        1   4   6   4   1
```

Procedures:
1. Each member chooses a role. The Triangle Maker writes six more rows of the triangle in the space above.

2. The Diagonal Selector draws a ring around a part of a diagonal of the triangle, including a 1. The Sum Finder finds the sum of the circled numbers.

3. Procedure 2 is repeated for five more diagonals.

4. Discuss how to find the sum of any diagonal without adding by examining the numbers in the triangle. The Sum Finder writes a summary of the discussion.

_____

_____

5. The Triangle Maker makes four drawings of the triangular numbers that are formed by adding the natural numbers: 1, 1 + 2 = 3, 1 + 2 + 3 = 6, and so on.

6. The Sum Finder finds the sum of the first ten triangular numbers. What number in Pascal's triangle corresponds to this sum?

_____

NAME _____ CLASS _____ DATE _____

Cooperative-Learning Activity
11.8 Coefficient Challenge

Group members: 3

Materials: calculators

Roles: **Selector** chooses a combination that is to be a coefficient in a binomial expansion

Polynomial Identifier writes a power of a binomial in the form $(a + b)^n$, identifies n, and the specific term that has the coefficient chosen by the Selector

Checker checks the work of the Polynomial Identifier

Preparation: You will participate in a contest in which you try to achieve the highest score. Each member keeps track of his or her own score.

Procedures:
1. Each member chooses a role.

2. The Selector chooses a combination, such as $\binom{8}{4}$, and gives it to the Polynomial Identifier.

3. The Polynomial Identifier writes a polynomial in the form $(a + b)^n$ and writes the specific term that has the given coefficient in terms of factorial notation and powers of a and b.

Polynomial	Specific term	Player/Score

4. The Checker checks the work. If it is correct, the Polynomial Identifier adds 1 point to his or her score. If it is incorrect, 1 is subtracted from the score.

5. Procedures 2–4 are repeated for seven more coefficients.

6. Rotate roles twice and repeat Procedures 2–5. The member with the highest score is the winner. Another round is played if there is a tie.

Algebra 2   Cooperative-Learning Activity   75

NAME _____ CLASS _____ DATE _____

Cooperative-Learning Activity
12.1 Measures of Central Tendency Game

Group Members: 2

Materials: penny, 32 square tiles, calculator

Responsibilities: Prepare and play a game.

Preparation: Imagine that a ball is rolled through the triangular array of pegs shown in the diagram at left below. The path that the ball takes as it moves through the array will be determined by the directional change caused by contact with any of the pegs. The ball has an equal chance of going to the left or to the right with each contact. As a result, it can finish at any of the 6 bins. One possible path that a ball might travel is shown in the diagram.

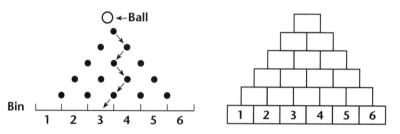

You will use a tile to represent the ball and the game board at right above to represent the array of pegs. A coin toss will determine which direction the ball will go. If the coin shows heads, the ball goes right. If the coin shows tails, the ball goes left.

Procedures:
1. Enlarge the game board shown at right above.

2. You will move 32 tiles through the game board by using a coin toss, as described above. Before playing the game, make a prediction of how many tiles will end up at each finish bin based on your knowledge of probability. Record your predictions in a frequency table created by the group. Then find the mean, median, mode, and range of the data you predicted.

3. Play the first 16 tiles with the first player tossing the coin while the second player moves the tile. Then play the last 16 tiles with the second player tossing the coin while the first player moves the tiles.

4. After each trial, make a tally mark representing the finish bin. After the 32nd trial, record the total for each bin in a frequency table created by the group. Then calculate the mean, median, mode, and range of the actual data and compare the experimental result with your predictions. Describe what you notice.

_____

_____

_____

NAME _____ CLASS _____ DATE _____

Cooperative-Learning Activity
12.2 Choosing a Display for a Data Set

Group members: 3

Materials: calculator

Responsibilities: You will analyze data about world population and decide how to best display the changes that occurred between 1950 and 1995.

Preparation: The world populations by regions are given for the years 1950 and 1997 in millions.

Year	North America	Latin America	Europe	Asia*	Africa	Oceania	Total
1950	166	164	392	1560	219	13	2513
1997	297	496	508	3770	750	29	5852

* including the former U.S.S.R.

Procedures:

1. Discuss different types of displays that might be used to represent the data. Display types include stem-and-leaf plots, histograms, and circle graphs. Choose the display that you think is best suited to this data, and explain your choice.

   _____

   _____

2. What information must be calculated in order to complete the display that you chose? Work together to find the information.

   _____

   _____

   _____

3. Divide the work and make the display to show the 1950 and 1997 world populations.

4. What changes that occurred between 1950 and 1997 can you identify from your display?

   _____

Algebra 2   Cooperative-Learning Activity   **77**

NAME _____ CLASS _____ DATE _____

Cooperative-Learning Activity
12.3 Testing Reading Passages

Group members: 2

Materials: no special materials

Roles: **Word Counter** counts the number of letters in the words of passages, not counting any punctuation

Box-and-Whisker Plotter makes a box-and-whisker plot

Preparation: One way to determine if a reading passage is suitable for different-aged children is to count the lengths of the words in the passage.

Excerpt A: It's magic. You've seen it done. Your uncle has a favorite card trick, or your friend can make a coin disappear. They make it look so easy. But did you know that it takes lots of practice to make a trick look that simple?

(From *Better Homes and Gardens*, October 1997)

Excerpt B: The Gradgrind party wanted assistance in cutting the throats of the Graces. They went about recruiting, and where could they enlist recruits more hopefully than among the fine gentlemen who, having found out everything to be worth nothing, were equally ready for anything?

(From *Hard Times For These Times* by Charles Dickens)

Procedures: 1. Each member chooses a role.

2. The Word Counter uses the space below to tally the number of letters for each word of excerpt A.

3. The Box-and-Whisker Plotter makes a box-and-whisker plot of the data.

4. Switch roles and repeat Procedures 2 and 3 for excerpt B.

5. Discuss the two box-and-whisker plots in terms of suitability for children. Which excerpt is more appropriate for younger children? What other factors must be considered?

_____

_____

_____

NAME _____ CLASS _____ DATE _____

Cooperative-Learning Activity
12.4 Quality Assurance

Group members: 3

Materials: calculators

Responsibilities: You will calculate the mean deviation for a set of data to see if it satisfies the requirements of the Quality Control department.

Preparation: The Quality Control department of a die manufacturer will not release a batch of gauges if the mean deviation for the diameter is greater than 0.001 inches. The Production department has given your group the following diameters for 30 gauges whose diameter is supposed to be 0.75 inches.

0.750, 0.752, 0.751, 0.692, 0.698, 0.746, 0.749, 0.753, 0.750, 0.745,
0.748, 0.751, 0.753, 0.755, 0.748, 0.746, 0.749, 0.754, 0.750, 0.751,
0.752, 0.753, 0.750, 0.748, 0.747, 0.749, 0.750, 0.752, 0.753, 0.748

Procedures:

1. Calculate the mean of the data. _____

2. Divide the data into three groups of ten data values. Each member completes one of the tables below.

x_i	$|x_i - 0.75|$
Total	

x_i	$|x_i - 0.75|$
Total	

x_i	$|x_i - 0.75|$
Total	

3. Members each calculate the mean deviation (m.d.) for their data sets.

m.d. [ ]   m.d. [ ]   m.d. [ ]

What do the mean deviations separately and together tell you about the acceptability of the batch?

_____

Algebra 2

Cooperative-Learning Activity
12.5 Making the Presentation

Group members: 3

Materials: calculator

Roles: **Success Expert** calculates the probability of being chosen

Outcome Counter finds the coefficient in the Binomial Theorem

Binomial Theorem Expert writes an expression to represent the probability that a group will be selected at least three times in the next two weeks

Preparation: Mrs. Math has divided her class into seven cooperative groups that stay the same throughout the year. Each day she randomly selects one group to make a presentation to the class. She does not take into account whether a group has presented before. You will calculate the probability that a particular group will be selected at least three times in the next two weeks (10 days).

Procedures:
1. Each member chooses a role. Discuss why this situation is an example of a binomial experiment. Identify n and the successful outcomes.

   _____

   _____

2. The Success Expert calculates the probability, p, of success for a particular group to be chosen on any given day. Then he or she calculates the probability, q, that the group will not be chosen on any given day.

   _____

3. The Outcome Counter and the Binomial Theorem Expert write an expression to represent the probability that a particular group will be chosen exactly 3 times.

   _____

4. The Success Expert checks to see if the expression above is correct.

5. All the members calculate the probability in Procedure 3 to four decimal places and compare their answers. If there are different answers, decide which answer is correct.

   P(chosen exactly 3 times) _____

6. Discuss the most efficient way to find the probability of at least 3 selections and then work together to find this probability.

   _____

   _____

NAME _____ CLASS _____ DATE _____

Cooperative-Learning Activity
12.6 Ordering Cereal

Group members: 2

Materials: no special materials

Roles: **Statistician** sketches a curve that models a situation.

Store Manager determines how much cereal to order

Preparation: The manager of a grocery store has determined that the weekly demand for a popular cereal is normally distributed with a mean of 800 boxes and a standard deviation of 75 boxes. The manager wishes to order enough cereal per week so that there is only a two percent chance of running out of that cereal in one week.

Procedures:
1. Each member chooses a role.

2. In the space below, the Statistician uses the given information to sketch and label a normal curve to represent the sales of the cereal, showing the values for ±1, ±2, and ±3 standard deviations.

3. Use the graph to find the number of boxes that should be ordered with the following chances of running out:

   a. 16% _____

   b. 50% _____

   c. 84% _____

   d. 98% _____

4. The Store Manager decides how much to order so that there is a 2% chance of running out.

   _____

5. Discuss what the Store Manager should do if the cereal is going to be an advertised special for the week.

   _____

Algebra 2   Cooperative-Learning Activity   **81**

NAME _____ CLASS _____ DATE _____

Cooperative-Learning Activity
13.1 Solving a Surveying Problem

Group members: 3

Materials: calculator

Roles: **Modeler** sketches and labels a diagram

Equation Writer writes equations that model the problem

Solver solves the problem

Preparation: A surveyor wants to find the height of an ice formation protruding from a lake. She determines that when she is standing at the edge of the water, the angle from the ground to the tip of the ice is 40°. She then paces off 35 feet and determines that the angle from the ground to the tip of the ice is now 20°. You will determine the height of the ice above the water.

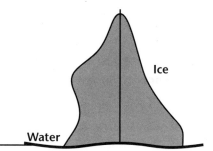

Procedures:
1. Each member chooses a role.

2. The Modeler adds appropriate lines to the drawing in order to form two right triangles. These right triangles both have the height of the ice as one leg and the line of sight as a hypotenuse. Then the Modeler adds the known information to the diagram and assigns a variable name to the height of the ice.

3. Using the information in the diagram, the Equation Writer writes two equations involving trigonometric ratios.

   Equation 1: _____

   Equation 2: _____

4. The Solver solves the equations to determine the height of the ice formation.

   _____

5. Discuss how this surveying information, diagram, and solution could be used in a different situation.

   _____

   _____

82 Cooperative-Learning Activity   Algebra 2

NAME _____ CLASS _____ DATE _____

Cooperative-Learning Activity
13.2 Playing a Trigonometric Function Game

Group members: 2

Materials: 10 index cards

Responsibilities: You will play a game in which you turn over cards to determine a quadrant and the value of a trigonometric function for some angle in that quadrant. You will find the values of the six trigonometric functions for the angle that meets the specifications indicated by the quadrant card and the value card.

Preparation: Prepare the cards by writing one of the numerals I, II, III, and IV on four of the cards. On each of the other cards, write one of the following:

$$\left|\sin\theta\right| = \frac{\sqrt{2}}{2}, \quad \left|\cos\theta\right| = \frac{\sqrt{3}}{2}, \quad \left|\tan\theta\right| = \frac{7}{24},$$
$$\left|\csc\theta\right| = \frac{13}{12}, \quad \left|\sec\theta\right| = 4, \quad \left|\cot\theta\right| = \frac{3}{4}$$

Keep the stack of four cards separate from the stack of six cards.

Procedures:
1. One of the members shuffles each stack of cards and turns over the top card on each stack. The same member determines the **sign** of each of the six trigonometric functions determined by the quadrant and the value.

   For example, if the cards show II and $\left|\cos\theta\right| = \frac{\sqrt{3}}{2}$, this member would write

   $$\sin\theta = + \qquad \cos\theta = -\frac{\sqrt{3}}{2} \qquad \tan\theta = -$$
   $$\csc\theta = + \qquad \sec\theta = - \qquad \cot\theta = -$$

2. The other member completes the values. If the sign is wrong, he or she may change the sign before proceeding. The first member checks the answers and gives one point for each correct answer. Each member keeps track of his or her score.

$\sin\theta =$	$\cos\theta =$	$\tan\theta =$
$\csc\theta =$	$\sec\theta =$	$\cot\theta =$

   SCORE: _____

3. Switch roles and repeat Procedures 1 and 2.

$\sin\theta =$	$\cos\theta =$	$\tan\theta =$
$\csc\theta =$	$\sec\theta =$	$\cot\theta =$

   SCORE: _____

4. Continue the game until one member has a total of 18 points or more.

Algebra 2

NAME _____ CLASS _____ DATE _____

Cooperative-Learning Activity
13.3 Continuing Piecewise and Periodic Functions

Group members: 2

Materials: no special materials

Roles: Function Maker creates a piecewise function whose stated domain is an interval
Function Grapher graphs the given function over the stated interval and continues the graph over more intervals

Preparation: A function is a piecewise function over an interval if its function values are given by different rules on different parts of the interval. A function is periodic if its values repeat in a patterned way. One cycle of the graph of a periodic function is a complete picture of the part of the graph that repeats.

Procedures:

1. Each member chooses a role.

2. The Function Maker creates a piecewise function over the interval $0 \le x < 4$, such as the one shown at right.

   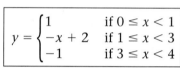
   $$y = \begin{cases} 1 & \text{if } 0 \le x < 1 \\ -x + 2 & \text{if } 1 \le x < 3 \\ -1 & \text{if } 3 \le x < 4 \end{cases}$$

3. The Function Grapher sketches a graph of the function over the interval $0 \le x < 4$. He or she then sketches a continuation of the graph over the intervals $4 \le x < 8$, $8 \le x < 12$, and $12 \le x < 16$ on the same grid.

   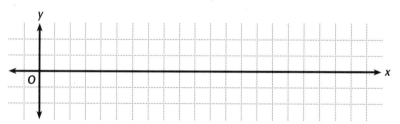

4. Switch roles. Now assume that the piecewise function represents one cycle of a periodic function. The Function Grapher chooses the period of the function, and graphs the function along with several cycles of its graph.

   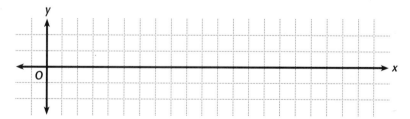

5. Discuss how to determine whether a function is periodic by examining its graph.

   _____

   _____

NAME _____ CLASS _____ DATE _____

Cooperative-Learning Activity
13.4 Radians and Degrees

Group members: 2

Materials: no special materials

Roles: **Converter** changes degree measure to radian measure and vice versa

Length Finder uses proportion to find arc length

Preparation: You will use proportions to go back and forth between degree and radian measures, and you will find the arc lengths intercepted by the given angle.

Procedures: 1. Each member chooses a role. Complete the following table.

	Central angle	Radius of circle	Radian measure	Arc length
a.	180°	2 inches		
b.	90°	2 inches		
c.	45°	2 inches		
d.	30°	2 inches		

2. Switch roles and complete the following table.

	Central angle	Radius of circle	Radian measure	Arc length
a.	180°	4 inches		
b.	90°	4 inches		
c.	45°	4 inches		
d.	30°	4 inches		

3. Compare the answers to Procedures 1 and 2 and discuss any relationships that you see. The last Converter writes a summary.

_____

4. Switch roles again and complete the following table.

	Central angle	Radius of circle	Degree measure	Arc length
a.	$\frac{3\pi}{2}$	3 inches		
b.	$\frac{7\pi}{4}$	2 inches		
c.	$\frac{7\pi}{6}$	5 inches		

5. Can degree measure be used to find arc length? Explain.

_____

Algebra 2   Cooperative-Learning Activity   **85**

NAME _____ CLASS _____ DATE _____

Cooperative-Learning Activity
13.5 Finding Periods and Phase Shifts

Group members: 4

Materials: no special materials

Roles: **Director** identifies transformations of the sine or cosine function from the parent function

Period and Amplitude Calculator finds the period of the function

Phase Calculator identifies phase shift of the given function

Grapher sketches the graph using information provided by other members

Preparation: You will identify and sketch the graphs of trigonometric functions that are stretches, compressions, and horizontal translations of their parent functions. You will use the following functions:

A. $y = 3 \sin x$
B. $y = \sin(3x)$
C. $y = \sin\left(x + \frac{\pi}{3}\right)$
D. $y = \cos\left(\frac{1}{2}x\right)$

Procedures:
1. Each member chooses a role.

2. The Director identifies function A as having a vertical stretch or compression, a horizontal stretch or compression, or a horizontal translation of a parent trigonometric function.

   _____

3. The Period and Amplitude Calculator and the Phase Calculator determine the amplitude, period, and phase shift as appropriate.

   _____

4. The Grapher uses the information provided in Procedure 3 to sketch the graph.

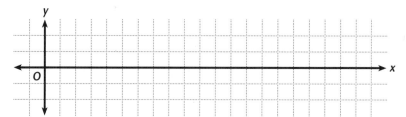

5. Repeat Procedures 2–4 for functions B–D, using the grid below.

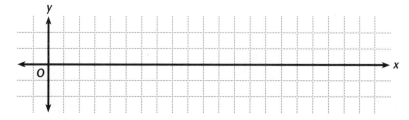

Cooperative-Learning Activity
13.6 Casting Shadows Experiment

Group members: 4

Materials: flashlight, cylindrical object, ruler

Roles: **Object Holder** holds the object perpendicular to a desk or table

Light Holder holds the flashlight at different heights

Measurer measures the shadow and the object

Angle Calculator finds the angle of depression, θ

Preparation: You will examine the relationship between angle of elevation and shadow length. One member will hold an object vertically while the other moves a flashlight to create shadows. It is advisable to turn off the room lights.

Procedures:
1. Each member chooses a role.

2. The Object Holder holds the cylindrical object stationary throughout the experiment. The Measurer measures and records the height of the cylindrical object. The Light Holder shines the flashlight from a position one foot to the right and slightly above the object.

3. The Measurer records the object's height and the length of the shadow. The Angle Calculator finds the angle of depression for the flashlight. The second column of the table below will contain the measurements and calculations.

Object height						
Length of shadow						
Angle of depression						

4. Repeat Procedures 2 and 3 five times. Each time the Light Holder changes the angle of depression for the flashlight.

5. Discuss the relationship between the length of the shadow and the angle of depression. The Object Holder writes a summary of the discussion.

6. Discuss how this experiment models the relationship between the movement of the Sun and lengths of shadows.

Algebra 2

NAME _____ CLASS _____ DATE _____

Cooperative-Learning Activity
14.1 Using the Law of Sines

Group members: 3

Materials: yardstick, protractor, 2 desks, calculator

Roles: **Data Gatherer** measures the distance between the two desks and measures the angles

Modeler writes equations by using the law of sines to find distances

Solver solves the equations to find the distances

Preparation: You will use the law of sines to calculate the distance between a student's desk at A and the teacher's desk or podium at C and the distance between the podium and another student's desk at B.

Procedures: 1. Each member chooses a role.

2. Select two desks in the classroom.

3. The Data Collector measures and records the distance between the desks, represented by A and B, to the nearest inch. To measure the angles, the Data Collector uses the yardstick and another straight object to form $\angle A$ and $\angle B$, as shown in the diagram above. Those angle measures are also recorded in the table.

AB	m∠A	m∠B

4. The Modeler uses the law of sines to write two equations whose solutions will give AC and BC.

Equation for AC _____

Equation for BC _____

5. The Solver solves the equations to find the distances, to the nearest tenth of an inch, between each student desk and the teacher's desk. Check that these distances are correct by actually measuring AB and BC.

$AC =$ _____ $BC =$ _____

6. Change the measurements as shown in the table below, and recalculate AC and BC.

AB	m∠A	m∠B
2 inches longer than before	2° more than before	3° less than before

$AC =$ _____ $BC =$ _____

88 Cooperative-Learning Activity   Algebra 2

NAME _____ CLASS _____ DATE _____

Cooperative-Learning Activity
14.2 Solving Triangles

Group members: 3

Materials: calculators

Roles: Director labels a triangle and decides best approach for solving the triangle

Law-of-Sines Expert uses the law of sines to find unknown parts of triangles

Law-of-Cosines Expert uses the law of cosines to find unknown parts of triangles

Preparation: You will solve each of the following triangles by using the law of sines, the law of cosines, or both. You will find all solutions if more than one triangle exists.

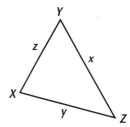

A. $y = 6, z = 9, m\angle X = 30°$
B. $x = 9, m\angle Y = 50°, m\angle Z = 70°$
C. $m\angle Y = 20°, x = 4, y = 2$
D. $x = 4, y = 3, z = 9$

Procedures:
1. Each member chooses a role.

2. Using pencil, the Director labels the triangle above with the information about triangle A. The other members review the labeling and discuss whether other triangles are possible. Describe the given information as SSS, SSA, SAA, ASA, or SAS information.

   _____

3. Discuss two approaches to solving the triangle. Then find the solution.

   _____

   _____

4. Switch roles and repeat Procedures 2 and 3 to solve triangle B.

   B: Information type: _____   Solution: _____

5. Switch roles again to solve triangles C and D.

   C: Information type: _____   Solution: _____

   D: Information type: _____   Solution: _____

6. Discuss how to solve a triangle problem in various ways.

   _____

Algebra 2 Cooperative-Learning Activity **89**

NAME _____ CLASS _____ DATE _____

Cooperative-Learning Activity
14.3 Simplifying Trigonometric Expressions

Group members: 2

Materials: no special materials

Roles: **Sine/Cosine Writer** writes expressions in terms of sines and cosines only

Single-Function Writer writes expression in terms of one other trigonometric function

Preparation: The first column of the table below contains six trigonometric expressions. The second column gives the expression rewritten in terms of sine and cosine. The third column gives the expression in terms of one other trigonometric function.

	Expression	Written in terms of sin and cos	Written in terms of one other function
Example	$\sec^2 \theta - 1$	$\dfrac{1}{\cos^2 \theta} - 1$	$\tan^2 \theta$
A.	$\tan \theta$		
B.	$\dfrac{\sec \theta}{\csc \theta}$		
C.	$1 - \sin^2 \theta$	—	
D.	$\cot \theta \sec \theta$		
E.	$\tan^2 \theta + 1$		
F.	$\dfrac{\tan \theta}{\cot \theta}$		

Procedures:
1. Each member chooses a role.

2. The Sine/Cosine Writer and Single-Function Writer complete row A.

3. Switch roles and complete row B.

4. Switch roles four more times and complete rows C, D, E, and F.

5. Discuss why rewriting an expression in terms of sine and cosine only or rewriting it in terms of one other function might be helpful strategies when proving an identity or solving an equation.

_____

_____

_____

NAME _____ CLASS _____ DATE _____

Cooperative-Learning Activity
14.4 Using Sin 1° and Cos 1° to Evaluate Sine and Cosine

Group members: 2

Materials: calculators

Responsibilities: You will generate a table of sines and cosines for whole-number degrees.

Preparation: To make the table, you will use the values of sine and cosine given below.

$$\sin 1° \approx 0.0175 \quad \text{and} \quad \cos 1° \approx 0.9998$$

You will also use the addition formulas given below.

$$\sin(A + B) = \sin A \cos B + \sin B \cos A$$

$$\cos(A + B) = \cos A \cos B - \sin A \sin B$$

Procedures:

1. One member uses the given values of sin 1° and cos 1°, along with the addition formulas, to complete the column headed sin $n°$. The other member does the same to complete the column headed cos $n°$. Calculated values should be rounded to the nearest ten-thousandth before completing the next row.

$n°$	sin $n°$	sin $n°$	cos $n°$	cos $n°$
$n° = 2°$	sin (1° + 1°)		cos (1° + 1°)	
$n° = 3°$	sin (2° + 1°)		cos (2° + 1°)	
$n° = 4°$	sin (3° + 1°)		cos (3° + 1°)	
$n° = 5°$	sin (4° + 1°)		cos (4° + 1°)	

2. Discuss how to continue the table of sines and cosines. In particular, describe how generating the table is related to sequences that are defined recursively.

_____

_____

_____

3. Discuss how the table of values diverges from the actual values of sine and cosine as n increases, and offer an explanation of this divergence.

_____

_____

4. The table is to be expanded to include sines and cosines of degree measures that are whole numbers and whole numbers plus one half of one degree. What additional information is need to do this?

_____

Algebra 2   **Cooperative-Learning Activity**

NAME _____ CLASS _____ DATE _____

Cooperative-Learning Activity
14.5 Exploring Roof Systems

Group members: 2

Materials: no special materials

Roles: **Architect** makes specifications for a roof system
Mathematician uses trigonometric equations to answer roof-system questions

Preparation: You will explore variations in roof system specifications by using a half-angle identity for the sine function.

Procedures:
1. Each member chooses a role.

2.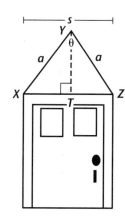
   a. The Mathematician uses right-triangle trigonometry and the diagram at right to write an equation that relates $\frac{\theta}{2}$, a, and s, and that involves the sine function. _____
   b. The Mathematician rewrites the equation from part **a** using the half-angle identity for the sine function to write an equation relating θ, a, and s. _____
   c. The Mathematician explains why + was used in part **b** rather than ±.

   _____

3. The Architect specifies the roof angle, θ, and the doorway width, s. _____

4. The Mathematician uses the information provided by the Architect to find a. _____

5. Switch roles and repeat Procedures 3 and 4.

   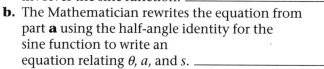
   $\theta =$ _____ $s =$ _____ $a =$ _____

6. In a roof system, $0° \leq \theta \leq 180°$. Work together to solve the equation written in part **b** of Procedure 2 for θ in terms of a and s. How is an inverse trigonometric function used in the solution?

   _____

   _____

   _____

7. Explain how to use the equation written in Procedure 6.

   _____

NAME _____ CLASS _____ DATE _____

Cooperative-Learning Activity
14.6 Exploring Distance for a Kicked Football

Group members: 2

Materials: calculators

Roles: **Modeler** sketches and labels a diagram
Solver solves the equation

Preparation: The horizontal distance, d, in feet that a football is kicked can be modeled by the trigonometric equation below.

$$d = \left(\frac{1}{32}\right)(v_0)^2 \cdot \sin 2\theta,$$

where θ is the angle made between the horizontal and the path of the kick, and v_0 is the initial velocity in feet per second of the ball when it is kicked.

You will solve a trigonometric equation to answer the following question.

A football is kicked with an initial velocity of 82 feet per second from the west 40-yard line. A receiver on the east 10-yard line catches the ball. At what angle was the ball kicked?

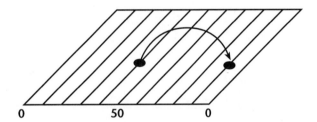

Procedures:
1. Each member chooses a role.

2. The Modeler sketches and labels a diagram to represent the situation.

3. The Solver solves the equation. _____

4. How would the angle change if the kick were shorter and the initial velocity stayed the same?

   _____

   _____

5. Test your conclusion from Procedure 4 by finding the angle when the kick from the west 40-yard line lands at the east 20-yard line.

   _____

6. Discuss how to change the kick angle so that the football travels the greatest distance for a given initial velocity.

   _____

   _____

Algebra 2   Cooperative-Learning Activity   **93**

ANSWERS

Cooperative-Learning Activity — Chapter 1

Lesson 1.1

1–4. Answers may vary. Sample answer:

Fixed cost	Variable cost	Total cost
$1000	$8	$1008
$1000	$16	$1016
$1000	$24	$1024
$1000	$32	$1032
$1000	$40	$1040
$1000	$48	$1048
$1000	$56	$1056

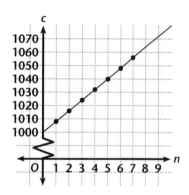

5. $C = 1000 + 8n$

6. $1840, $1960, $2040, $2136

Lesson 1.3

4–6. Answers may vary. Sample answer:

a	Number of solutions
10	11
7	8
4	5
5	6

7. The number of ordered pairs is one more than a.

8. Answers may vary. Sample answer: For each number up to and including a, there is a coordinate pair with that number as an x-coordinate. For example, for $a = 2$, you have (0, 2), (1, 1), and (2, 0), so there are three coordinate pairs.

Lesson 1.4

2.

Radius	Estimate of circumference
1 inch	6.2 inches
2 inches	12.4 inches
4 inches	25.12 inches

3. $R = \frac{C}{6.2}, \frac{1}{6.2}$

4.

Radius	Fraction of circle
1 inch	1
2 inches	$\frac{1}{4}$
4 inches	$\frac{1}{16}$

5. Answers may vary. Sample answer: If the radius of a circle is r, the area of the circle is r^2 times the area of the circle with radius 1.

Lesson 1.5

Answers may vary. Sample answers:

2. There may be a negative correlation between the number of people who go to the movies and age group.

3. (1, 70), (2, 68), (3, 58), (4, 50)

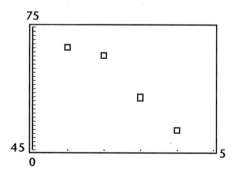

4. negative

ANSWERS

5. $y = -7x + 79$

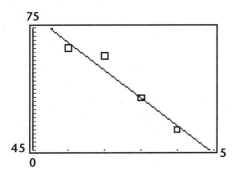

6. All the relationships between the age groups exhibit a negative correlation except for gardening, which has a strong positive correlation with age.

Lesson 1.6

2. a. $x = 4$
   b. $x = 2\frac{1}{2}$
   c. $x = \frac{1}{3}$
   d. $x = -2$

3. a. $x = 11\frac{1}{5}$
   b. $x = -39$
   c. $x = -12\frac{3}{4}$
   d. $x = 80$

Lesson 1.7

2. Let x represent the extra amount per child than can be set aside each month. Then $3x + 300 \leq 570$.

3. at most $90

4. They could have as little as no money left and as much as $1000 left for additional expenses.

5. According to one contractor, labor costs will be between $18,000 and $24,000. According to the other contractor, labor costs will be between $15,000 and $22,000.

Lesson 1.8

2.

	Top	Middle	Bottom
a.	$-10 \leq x \leq -8$	$-4 \leq x \leq -2$	$-7 \leq x \leq -5$
b.	$5 \leq x \leq 7$	$-1 \leq x \leq \frac{1}{2}$	$2 \leq x \leq 4$
c.	$2 \leq x \leq 4$	$\begin{cases} 5 < x < 7 \\ 8 < x < 10 \end{cases}$	$-4 \leq x \leq -2$
d.	$\begin{cases} -4 \leq x \leq -1 \\ 0 \leq x \leq 1 \end{cases}$	$-10 \leq x \leq -8$	$-1 < x < 1$
e.	$8 \leq x \leq 10$	$2 \leq x \leq 4$	

3. The three number lines spell ALGEBRA in block letters.
   Answers may vary. Sample answer: Absolute value inequalities must be split into 2 inequalities. In order to solve an inequality with *and*, the solution includes all the portions of the number line that satisfy all the inequalities.

Cooperative-Learning Activity — Chapter 2

Lesson 2.1

3. Answers may vary. Sample answer:

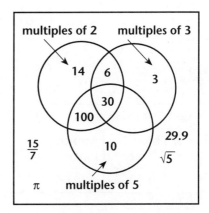

5. The number must be a multiple of 2, 3, and 5; the number is a multiple of 2 and 5 and might be a multiple of 3; the number is not a multiple of 2 or 5 and might be a multiple of 3.

ANSWERS

Lesson 2.2

1.

1	1	1	$\frac{1}{8x^5y^8}$	$2^{\frac{2}{3}}$
$2x^3y^5$	$\frac{1}{8x^5y^8}$	1	$2^{\frac{2}{3}}$	$\frac{1}{8x^5y^8}$
1	$2x^3y^5$	$2^{\frac{2}{3}}$	1	$\frac{1}{8x^5y^8}$
1	$2x^3y^5$	$2^{\frac{2}{3}}$	$2x^3y^5$	$\frac{1}{8x^5y^8}$

2. $1 = \left(\frac{5x^0y^5}{25^{\frac{1}{2}}y^5}\right) = \left(-\frac{16}{24}\right)^0 = (-4^{\frac{1}{3}})^0 = (4^0)^{\frac{2}{6}} =$

$(2^0)^{\frac{1}{9}} = (2x^5y^8)^0; 2^{\frac{2}{3}} = \sqrt[3]{4} = (2^{\frac{1}{3}})^2 = \left(\frac{64}{16}\right)^{\frac{1}{3}};$

$\frac{1}{8x^5y^8} = \left(\frac{-8(-x)^5}{y^{-8}}\right)^{-1} = \frac{4^{-1}x^3y^{-5}}{2x^8y^3} =$

$\frac{(2x^5y^{10})^{-3}}{x^{-7}y^{-22}} = \left(\frac{2^3x^{10}y^{16}}{x^5y^8}\right)^{-1}; 2x^3y^5 =$

$\frac{3^0x^{-5}y^{-2}}{2^{-1}x^{-8}y^{-7}} = (8x^9y^{15})^{\frac{1}{3}} = \frac{4^{\frac{1}{2}}xy^{-2}}{8^0x^{-2}y^{-7}}$

3. A simplified expression is less complicated to work with, so you are less likely to make mistakes.

Lesson 2.3

2.

Option 1: 4 years at 8%		
Number of payments	Monthly rate	Monthly payment
48	0.0067	$703.63

Option 2: 3 years at 10%		
Number of payments	Monthly rate	Monthly payment
36	0.0083	$928.75

3.

Option 1: 4 years at 8%	
x	Balance due
1	$28,096.37
2	$27,392.74
3	$26,689.11
4	$25,985.48

Option 2: 3 years at 10%	
x	Balance due
1	$27,871.25
2	$26,942.50
3	$26,013.75
4	$25,085.00

4. Let y represent the balance due and x represent the number of payments made.
Option 1: $y = 28{,}800 - 703.63x$, where x is at most 48
Option 2: $y = 28{,}800 - 928.75x$, where x is at most 36

Lesson 2.4

2. Answers depend on value of x chosen. Sample answers for $x = 2$:
   a. -36
   b. -31
   c. 360
   d. 360
   e. $-1\frac{1}{8}$
   f. -5

3 a. $x^2 + x - 42$
  b. $-19x + 7$
  c. $20x^2 + 140x$
  d. $20x^2 + 140x$
  e. $\frac{x^2 - 49}{20x}$
  f. $\frac{x^2 - 49}{x + 7}$

ANSWERS

4. Answers depend on value of x chosen.
Sample answers for x = −1:
  a. −13
  b. 41
  c. 120
  d. −13

5. Addition and multiplication can be carried out in either order.

Lesson 2.5

2. a. Let w = amount of water and d = depth; $w = 800d$
  b. $d = \dfrac{w}{800}$
  c. The depth of the swimming pool is determined by dividing the amount of water by 800. The inverse relationship is a function.

3. a. Let A be the amount of material and s = width of table cloth
$A = s^2$
  b. $s = \sqrt{A}$
  c. The length of a side of a table cloth of A square units is the square root of A. Mathematically, the inverse is not a function because the solution to the equation is $\pm A$. However, if the domain is restricted to positive values for s (as in this example), the inverse is a function.

4. Situation D: Let R represent the retail price and S represent the sale price.
  a. $S = (0.7)^2 R = 0.49R$
  b. $R = \dfrac{S}{.49}$
  c. The retail price can be found by dividing the sales price by 49%. The inverse relationship is a function.

Lesson 2.6

1. Answers may vary. Sample answer:
$(-0.6, -1), (-0.2, 0), (-0.4, 0), (-0.8, -1), (-0.5, -1)$

2. −3, 5

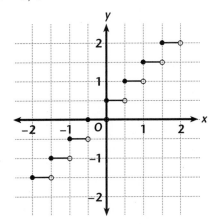

4. $y = 0$ if $-0.5 < x \le 0$
$y = -1$ if $-1 \le x \le -0.5$

6. $y = [x + 0.5]$

7. Yes; the rounding function is an example of a step function.

Lesson 2.7

3. a. $y = |x + 4|$ | $y = (x + 4)^2$ | $y = x + 4$
  b. $y = |x - 8| - 6$ | $y = (x - 8)^2 - 6$ | $y = x - 14$
  c. $y = |x + 2| + 5$ | $y = (x + 2)^2 + 5$ | $y = x + 7$

4. a. $y = |x + 4|$ | $y = (x + 4)^2$ | $y = x + 4$
  b. $y = |x - 4| - 6$ | $y = (x - 4)^2 - 6$ | $y = x - 10$
  c. $y = |x - 2| - 1$ | $y = (x - 2)^2 - 1$ | $y = x - 3$

5. Add the amounts used in the horizontal translations, keeping track of direction; then add the amounts used in the horizontal translations, keeping track of direction. Write one translation using the sums.

ANSWERS

Cooperative-Learning Activity — Chapter 3

Lesson 3.1

2. $0.05n + 0.10d = 0.75$

3–4. $d = 7.5 - 0.5n$

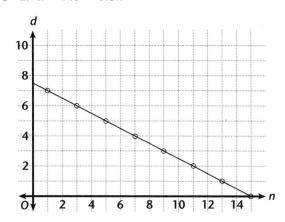

5. There are 8 possible answers to the question, "How many coins could you have in all?" The possible answers are 8, 9, 10, 11, 12, 13, 14, 15.

6. Answers may vary. The 4 systems should all be of the form
$$\begin{cases} 0.05n + 0.10d = 0.75 \\ n + d = t \end{cases}$$
where t is a number from 8 to 15.

Lesson 3.2

2. Answers may vary. The following is a sample set of systems of equations, using $x = 5$ and $y = 2$.

a. $\begin{cases} 3x - 2y = 11 \\ x + 4y = 13 \end{cases}$

b. $\begin{cases} 3x - 2y = 11 \\ 3x - 2y = 13 \end{cases}$

c. $\begin{cases} 3x - 2y = 11 \\ -6x + 4y = -22 \end{cases}$

5. Answers may vary. Sample answer: Inconsistent systems contain two equations whose graphs have the same slope and different y-intercepts. Dependent systems contain two equations where one equation is a multiple of the other.

Lesson 3.3

2.

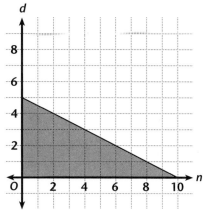

3. 36

4. 16, $0.5n + 0.10d \le 0.30$;
25, $0.5n + 0.10d \le 0.40$;

Lesson 3.4

1. Answers may vary. Sample answer: Let t = number of T-shirts and let c = number of CDs.

2. $c \ge 2t$
$15t + 18c \ge 270$

3–4.

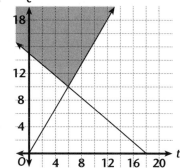

5. They should have at least 5 T-shirts and 10 CDs.

6. Answers may vary. Sample answer: There is not enough information to determine a maximum. The solution to the inequality is unlimited at the top of the region.

ANSWERS

Lesson 3.5

2. Answers may vary. Sample answer:

Inside

Coordinates	P
(1, 4)	−10
(3, 3)	−3
(3.5, 4.5)	−6.5
(3.5, 1.5)	2.5
(2, 4)	−8

Borders

Coordinates	P
(3, 2)	0
(2, 5)	−11
(4, 1.5)	3.5
(4, 2)	2
(0, 4.5)	−13.5

Vertices

Coordinates	P
(0, 4)	−12
(0, 5)	−15
(4, 5)	−7
$(4, \frac{4}{3})$	4

3. The maximum and minimum values always lie on the vertices. The maximum value is at $(4, \frac{4}{3})$, and the minimum value is at (0, 5).

Lesson 3.6

5. Situation A: $\begin{cases} x(t) = 6 - 2t \\ y(t) = -6 + t \end{cases}$

Situation B: $\begin{cases} x(t) = -6 + 2t \\ y(t) = 6 - 2t \end{cases}$

Cooperative Learning Activity — Chapter 4

Lesson 4.1

2. $A = \begin{bmatrix} 48 & 16 \\ 30 & 16 \\ 50 & 40 \\ 30 & 16 \end{bmatrix}$, $B = \begin{bmatrix} 60 & 25 \\ 40 & 30 \\ 50 & 40 \\ 15 & 30 \end{bmatrix}$,

$C = \begin{bmatrix} 45 & 30 \\ 20 & 15 \\ 15 & 10 \\ 16 & 10 \end{bmatrix}$

3. $\begin{bmatrix} 153 & 71 \\ 90 & 61 \\ 115 & 90 \\ 61 & 56 \end{bmatrix}$

4. $H = \begin{bmatrix} 24 & 8 \\ 15 & 8 \\ 25 & 20 \\ 15 & 8 \end{bmatrix}$

5. $A' = A - H, B' = B, C' = C + H$

6. $A' = \begin{bmatrix} 24 & 8 \\ 15 & 8 \\ 25 & 20 \\ 15 & 8 \end{bmatrix}$, $B' = \begin{bmatrix} 60 & 25 \\ 40 & 30 \\ 50 & 40 \\ 15 & 30 \end{bmatrix}$,

$C' = \begin{bmatrix} 69 & 38 \\ 35 & 23 \\ 40 & 30 \\ 31 & 18 \end{bmatrix}$

7. $\begin{bmatrix} 153 & 71 \\ 90 & 61 \\ 115 & 90 \\ 61 & 56 \end{bmatrix}$

The matrix is the same as the matrix in Procedure 3. This result should be expected because moving some inventory from one store to another does not change the total amount of inventory.

ANSWERS

Lesson 4.2

1. Answers may vary. Sample answer:
$$[200 \quad 100 \quad 200] \begin{bmatrix} 20 \\ 15 \\ 10 \end{bmatrix}$$ A

   Matrix A must be as shown. The entries in the other matrix must be multiples of 25 and must total 500.

2. Answer may vary. Sample answer: For matrix shown above, the total income is $7500.

3. Answer may vary. Sample answer: 200 $20 tickets, 200 $15 tickets, 100 $10 tickets

4. Answer may vary. Sample answer: Our first guess of 200 $20, 100 $15, and 200 $10 tickets produced too little income, by $500. By changing 100 of the $10 seats to $15 seats, we increased the income by the required $500.

Lesson 4.3

1. Answers may vary. Sample answer:

	27	G	20	N	13	U	6
A	26	H	19	O	12	V	5
B	25	I	18	P	11	W	4
C	24	J	17	Q	10	X	3
D	23	K	16	R	9	Y	2
E	22	L	15	S	8	Z	1
F	21	M	14	T	7	?	0

2. Answers may vary. Sample answer:
   ARE YOU SAD?
   A matrix consisting of
   $$\begin{bmatrix} 5 & -6 \\ -6 & 5 \end{bmatrix}$$
   multiplied by
   $$\begin{bmatrix} 26 & 22 & 2 & 6 & 8 & 23 \\ 9 & 27 & 12 & 27 & 26 & 0 \end{bmatrix}$$
   equals
   $$\begin{bmatrix} 76 & -52 & -62 & -132 & -116 & 115 \\ -111 & 3 & 48 & 99 & 82 & -138 \end{bmatrix}$$

3. $$\begin{bmatrix} -\frac{5}{11} & -\frac{6}{11} \\ -\frac{6}{11} & -\frac{5}{11} \end{bmatrix}$$

4. The decoded matrix should be the same as the first matrix in Procedure 2.

5. Answers may vary. Sample answer: NO
   $$\begin{bmatrix} 5 & -6 \\ -6 & 5 \end{bmatrix} \begin{bmatrix} 13 \\ 12 \end{bmatrix} = \begin{bmatrix} -7 \\ -18 \end{bmatrix}$$

6. The decoded matrix should be the same as the first matrix in Procedure 5.

Lesson 4.4

2.
	Trail mix 1	Trail mix 2	Mixture
No. of pounds	x	y	20
Amount of sunflower seeds	0.55x	0.20y	0.45(20)

3. $\begin{cases} x + y = 20 \\ 0.55x + 0.20y = 0.45(20) \end{cases}$

4. $14\frac{2}{7} \approx 14.3$ pounds of trail mix 1 and
   $5\frac{5}{7} \approx 5.7$ pounds of trail mix 2

5. $x + y = 20$
   $0.55x + 0.20y = 0.25(20)$
   $2\frac{6}{7} \approx 2.9$ pounds of trail mix 1 and
   $17\frac{1}{7} \approx 17.1$ pounds of trail mix 2

6. Answers may vary. Sample answer: Since 20 pounds of a mixture containing 45% sunflower seeds can be obtained using $14\frac{2}{7}$ pounds of trail mix 1 and a 55% mixture can be obtained using 20 pounds of trail mix 1 (that is, all trail mix 1), a 50% mixture can be obtained using $17\frac{1}{7} \approx 17.1$ pounds of trail mix 1. ($17\frac{1}{7}$ is midway between $14\frac{2}{7}$ and 20.) Subtract from 20 to find that you need $2\frac{6}{7} \approx 2.9$ pounds of trail mix 2.

ANSWERS

Lesson 4.5

2. Answers may vary. Sample answer:

$$R_1 \leftrightarrow R_3 \begin{bmatrix} 1 & -1 & 0 & 5 \\ -2 & 3 & 1 & 6 \\ 3 & 0 & -1 & 4 \end{bmatrix}$$

3. Steps may vary. Sample answer:

$$2R_1 + R_2 \to R_2 \begin{bmatrix} 1 & -1 & 0 & 5 \\ 0 & 1 & 1 & 16 \\ 3 & 0 & -1 & 4 \end{bmatrix}$$

$$-3R_1 + 3R_3 \to R_3 \begin{bmatrix} 1 & -1 & 0 & 5 \\ 0 & 1 & 1 & 16 \\ 0 & 3 & -1 & -11 \end{bmatrix}$$

$$R_2 + R_1 \to R_1 \begin{bmatrix} 1 & 0 & 1 & 21 \\ 0 & 1 & 1 & 16 \\ 0 & 3 & -1 & -11 \end{bmatrix}$$

$$-3R_2 + R_3 \to R_3 \begin{bmatrix} 1 & 0 & 1 & 21 \\ 0 & 1 & 1 & 16 \\ 0 & 0 & -4 & -59 \end{bmatrix}$$

$$-\tfrac{1}{4}R_3 \to R3 \begin{bmatrix} 1 & 0 & 1 & 21 \\ 0 & 1 & 1 & 16 \\ 0 & 0 & 1 & 14.75 \end{bmatrix}$$

$$-1R_3 + R_1 \to R_1 \begin{bmatrix} 1 & 0 & 0 & 6.25 \\ 0 & 1 & 1 & 16 \\ 0 & 0 & 1 & 14.75 \end{bmatrix}$$

$$-1R_3 + R_2 \to R_2 \begin{bmatrix} 1 & 0 & 0 & 6.25 \\ 0 & 1 & 0 & 1.25 \\ 0 & 0 & 1 & 14.75 \end{bmatrix}$$

4. The reduced row-echelon form is

$$\begin{bmatrix} 1 & 0 & \tfrac{2}{5} & 0 \\ 0 & 1 & -\tfrac{13}{5} & 0 \\ 0 & 0 & 0 & 1 \end{bmatrix}$$

5. The system is

$$\begin{cases} x - y + 3z = 2 \\ -2x + 2y - 6z = -1 \\ 4x + y - z = 2 \end{cases}$$

The third row of the reduced row-echelon form matrix shows that the system is inconsistent because it represents the equation $0x + 0y + 0z = 1$, or $0 = 1$.

Cooperative-Learning Activity — Chapter 5

Lesson 5.1

2.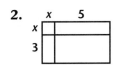

3. $x^2 + 5x + 3x + 15 = x^2 + 8x + 15$

4. Answers may vary. Sample answer:

x	Area
1	24
2	35
3	48
4	63
5	80

5. The minimum area is 15, which occurs when $x = 0$.

6.

```
    2x   1
  ┌────┬──┐
x │    │  │
  │    │  │
3 │    │  │
  └────┴──┘
```

$2x^2 + x + 6x + 3 = 2x^2 + 7x + 3$
Answers may vary. Sample answer:

x	Area
1	12
2	25
3	42
4	63
5	88

The minimum area is 3, which occurs when $x = 0$.

7. The area is $ax^2 + (b + ac)x + bc$.

Algebra 2

ANSWERS

Lesson 5.2

1. $BW = 4$, $BX = 13$, $YD = 8$, and $DZ = 17$

2. $WX = \sqrt{185}$, $YX = \sqrt{305}$, $YZ = \sqrt{353}$, and $WZ = \sqrt{233}$

3. Perimeter $= \sqrt{185} + \sqrt{305} + \sqrt{353} + \sqrt{233}$, about 65.12 units

4. Methods may vary. Sample method: Subtract the sum of the areas of the four triangles from the area of rectangle ABCD. The Pythagorean Theorem is not needed because only the lengths of the legs are need to find area.
Area $= 360 - (26 + 34 + 68 + 52) = 180$ square units

Lesson 5.3

2. Answers may vary. Sample answer: Addends should include a list of pairs of numbers whose sum is the number listed. The factors are all the number pairs whose product is given.
Sample answer for A:

Sum	Product
−11, 1	7, 3
−12, 2	−7, −3
−13, 3	21, 1
−5, −5	−21, −1
−6, −4	
−7, −3	
−8, −2	

3. A. −7, −3   F. −4, −2
   B. −8, 6   G. 12, −2
   C. 9, −8   H. 15, −2
   D. −4, −1   I. −10, −4
   E. −9, −3   J. −12, 3

4. Answers will vary. A sample list would look like the table on the worksheet.

5. Answers will vary. Check students' work.

Lesson 5.4

2. A. $\left(x - \left(-\frac{5}{2}\right)\right)^2 - \frac{1}{4} = 0$
   B. $\left(x - \frac{7}{2}\right)^2 - \frac{9}{4} = 0$
   C. $\left(x - \left(-\frac{11}{2}\right)\right)^2 - \frac{1}{4} = 0$
   D. $2\left(x - \left(-\frac{7}{4}\right)\right)^2 - \frac{169}{8} = 0$
   E. $3\left(x - \left(-\frac{19}{6}\right)\right)^2 - \frac{529}{12} = 0$
   F. $6\left(x - \frac{7}{12}\right)^2 - \frac{169}{24} = 0$

3. A. −2 and −3 B. 2 and 5 C. −6 and −5
   D. $\frac{3}{2}$ and −5 E. $\frac{2}{3}$ and −7 F. $-\frac{1}{2}$ and $\frac{5}{3}$

4. I. Write the given quadratic equation in the form $a(x - h)^2 + k = 0$.
   II. Solve $a(x - h)^2 + k = 0$ by using the Properties of Equality and taking a square root.

5. Step 1: $a(x - h)^2 = -k$, by the Addition Property of Equality
   Step 2: $(x - h)^2 = -\frac{k}{a}$ by the Multiplication Property of Equality
   Step 3: $x - h = \pm\sqrt{-\frac{k}{a}}$ by taking the square root of each side
   Step 4: $x = h \pm \sqrt{-\frac{k}{a}}$ by the Addition Property of Equality

Lesson 5.5

2.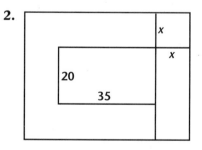

concrete: 804 square feet
tile: 502.5 square feet

3. Option 1: $(20 + 2x)(35 + 2x) - 700 = 804$, or $4x^2 + 110x - 804 = 0$
   Option 2: $(20 + 2x)(35 + 2x) - 700 = 502.5$, or $4x^2 + 110x - 502.5 = 0$

ANSWERS

4. Option 1: $x = 6$
   Option 2: $x \approx 4$

5. Answers may vary. Sample answer: The concrete should be used so that the patio area is larger.

Lesson 5.6

1. **a.** $7 + 6i$

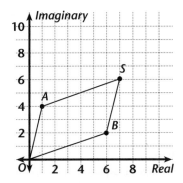

   b. $-4 + 5i$

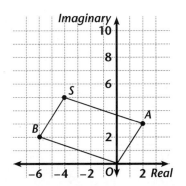

   c. $9 - 3i$

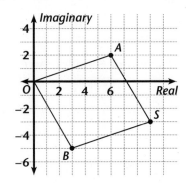

   d. $-4 - 5i$

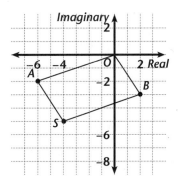

2. The quadrilateral formed by the origin, the points representing the given complex numbers, and the point representing their sum is a parallelogram.

3. Sketch the parallelogram with one vertex at the origin and two vertices at the given complex numbers. The fourth vertex gives the sum.

Lesson 5.7

1. **a.** first differences: $-1, -3, -5$
      second differences: $-2, -2$
      third differences: 0
   b. first differences: $6, 10, 14$
      second differences: $4, 4$
      third differences: 0
   c. first differences: $-2, -6, -10$
      second differences: $-4, -4$
      third differences: 0
   d. first differences: $4, 6, 8$
      second differences: $2, 2$
      third differences: 0

2. The first differences form a linear number pattern. The second differences are constant. The third differences are 0.

3. **a.** $P(n) = -n^2 + 2$
   b. $P(n) = 2n^2 - 5$
   c. $P(n) = -2n^2 + 7$
   d. $P(n) = n^2 + n + 2$

ANSWERS

Lesson 5.8

2. $C(x) = 75 + 20x$
$R(x) = (10 + x)(30 - x) = 300 + 20x - x^2$

3.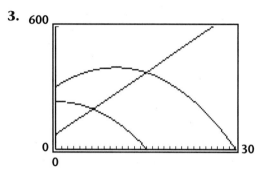

$P(x) = 225 - x^2$

4. $\begin{cases} y \geq 75 + 20x \\ y \leq 300 + 20x - x^2 \end{cases}$ Answers will vary.

Sample answer: The profit is 0 where the revenue and cost functions intersect, or at the x-intercept of the profit function. The profit will be 0 when 25 vases are sold (15 vases over the minimum).

5. Answers will vary. Sample answer:
By setting a minimum order of 20 vases and changing the reduction in price to $0.50 for each vase over the minimum, the students would make a profit for up to 53 vases (33 vases over the minimum). The revenue function would be $y = (20 + x)(30 - 0.5x)$.

Cooperative-Learning Activity — Chapter 6

Lesson 6.1

1.

	1990
Northeast	50,811,000
Midwest	59,667,000
South	85,773,000
West	52,782,000

2. Answers may vary. Sample answer:
The area with the smallest population changed from the West to the Northeast.

3. Answers may vary. Sample answer:

Ranking	2000	2010
1	South	West
2	West	South
3	Midwest	Northeast
4	Northeast	Midwest

4.

	2000	2010
Northeast	52,543,000	54,335,000
Midwest	60,478,000	61,301,000
South	97,610,000	111,080,000
West	64,531,000	78,896,000

5. Answers may vary. Sample answer:
The ranking of the regions did not change in the projected populations for the years 2000 and 2010.

Lesson 6.2

Answers may vary. Sample answer:

1–3.	Temperature in °C (x)	Time in seconds (y)
	2°	48
	20°	33
	30°	29
	38°	22
	45°	20

4–5. Answers may vary. Sample answer:

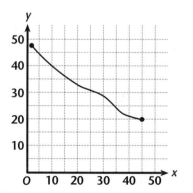

104 Answers   Algebra 2

ANSWERS

From this graph, the reaction will take approximately 23 seconds at a temperature halfway between that of the hot water and that of the tap water.

6. Answers may vary. Sample answer: Both are exponential functions, but the experimental graph decreases as the value of x increases, while the graph of $y = 2^x$ increases as the value of x increases.

7. Answers may vary. Sample answer: The rate of reaction is affected by the temperature of the water. The higher the temperature, the faster the reaction.

Lesson 6.3

2.

x	y
-4	$\frac{1}{16}$
-3	$\frac{1}{8}$
-2	$\frac{1}{4}$
-1	$\frac{1}{2}$
0	1
1	2
2	4
3	8

3–4.

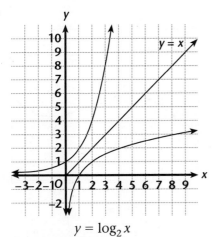

$y = \log_2 x$

5. Because these two graphs are reflections of each other across the line $y = x$, their functions are inverses.

Lesson 6.4

A. $\log_{10} 100 = x$, $10^x = 100$, $x = 2$

B. $\log_3 4 = x$, $2^x = 4$, $x = 2$

C. $\log_3 3x = 1$, $3^1 = 3x$, $x = 1$

D. $\log_8 \frac{1}{2} = x$, $8^x = \frac{1}{2}$, $x = -\frac{1}{3}$

E. $\log_3 \frac{81}{x} = 1$, $3^1 = \frac{81}{x}$, $x = 27$

Lesson 6.5

1. Answers may vary. Sample answer: Yes, it seems that twice the relative intensity would mean twice as loud. Twice as loud would mean 2 times the sound intensity.

2. Answers may vary. Sample answers:
Whisper = 20 decibels
City living room = 40 decibels

Whisper	City living room
$20 = 10 \log \frac{I}{I_0}$	$40 = 10 \log \frac{I}{I_0}$
$2 = \log \frac{I}{I_0}$	$4 = \log \frac{I}{I_0}$
$10^2 = \frac{I}{I_0}$	$10^4 = \frac{I}{I_0}$
$I = 100 I_0$	$I = 10{,}000 I_0$

Therefore the comparison of the intensity of a city living room to the intensity of a whisper is $10{,}000/100 = 100$. So the city living room is 100 times as loud.

3. Answers may vary. Sample answer: When the decibels are doubled, the sound intensity is increased by a factor of 10^2. This relationship is not linear.

Lesson 6.6

1. Answers may vary. Sample answer: The interest rate would need to be very large.

2. $1{,}000{,}000 = 5000e^{rt}$

3. Answers may vary. Sample answer:
Choosing $r = 10\%$, $t \approx 53$ years
Choosing $r = 8\%$, $t \approx 66$ years

ANSWERS

4. Answers may vary. Sample answer:
   For $r = 15\%$, $t \approx 35$ years
   For $r = 20\%$, $t \approx 26$ years

5. Answers may vary. Sample answer:
   These interest rates are very high. At a more realistic interest rate, it would probably take longer to reach $1,000,000.

Lesson 6.7

2. constant, $d = 18$; exponential, $c = \dfrac{\log 26}{\log 3}$, or about 2.97

3. linear, $a = \dfrac{4}{3}$; power equation, $b = \sqrt[3]{\dfrac{25}{3}}$, or about 2.03

4. If d, a, and b in $d = ab^c$ are known and c is unknown, then logarithms are required to solve the equation. The exception occurs when $\dfrac{d}{a}$ is a simple power of b.

5. Logarithms are needed when rate, initial deposit, and accumulation amount are known and time is unknown.

Cooperative-Learning Activity — Chapter 7

Lesson 7.1

1. Answers may vary. Sample answer:
   The 10-year annuity will have more value because the interest will be accumulating for a longer period of time.

2. Type 1:
   $500x^9 + 500x^8 + 500x^7 + 500x^6 + 500x^5 + 500x^4 + 500x^3 + 500x^2 + 500x + 500$

   $500(1.04)^9 + 500(1.04)^8 + 500(1.04)^7 + 500(1.04)^6 + 500(1.04)^5 + 500(1.04)^4 + 500(1.04)^3 + 500(1.04)^2 + 500(1.04) + 500$

   Type 2:
   $1000x^4 + 1000x^3 + 1000x^2 + 1000x + 1000$

   $1000(1.04)^4 + 1000(1.04)^3 + 1000(1.04)^2 + 1000(1.04) + 1000$

3. Plan 1: $6003
   Plan 2: $5416

4. Answers may vary. Sample answer:
   The guess was correct. The advantages of Type 1 are that the final amount is higher and that you can invest a smaller amount of money at one time. The advantage of Type 2 is that you acquire the money faster, even though the amount is $587 less.

5. Answers may vary. Sample answer:
   The insurance company could offer a higher interest rate for the shorter annuity. Keep Type 1 the same. Change Type 2 so that you invest $1000 at the beginning of each year for 5 years and earn 7% annual interest compounded yearly.
   Use $x = 1.07$ in the function for Type 2 above. Type 2 would have a final value of about $5751.

Lesson 7.2

1. Credit: $f = -2.59x^4 + 39.22x^3 - 177.59x^2 + 302.72x + 587.57$
   Unemployment:
   $f = 7.91x^4 - 32.57x^3 - 663.29x^2 + 4080.18x + 4190$

2.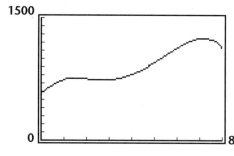

   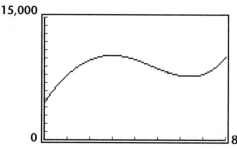

ANSWERS

3. Answers may vary. Sample answer: There is no obvious relationship. The fact that the outstanding credit is down in 1991 and 1992 could be related to the higher unemployment in those years, because purchasing of any kind would probably have been down.

Lesson 7.3

3. Answers may vary. Sample answer based on $a = -4$:

$$\begin{array}{r|rrrrr}
-4 & 2 & 3 & -2 & 7 & 6 \\
 & & -8 & 20 & -72 & 260 \\ \hline
 & 2 & -5 & 18 & -65 & \boxed{266}
\end{array}$$

4. $2x^3 - 5x^2 + 18x - 65$; remainder: 266

$(x + 4)\left(2x^3 - 5x^2 + 18x - 65 + \dfrac{266}{x+4}\right) = 2x^4 + 3x^3 - 2x^2 + 7x + 6$

Lesson 7.4

2. Answers may vary. Sample answer: 20 inches

3. $r \approx 5.2$ inches

4. For $h = 23$, $r \approx 4.9$ inches

5. Answers may vary. Sample answer: It is more practical to change the height because it is easier to hold a tank of a smaller radius on one's back. The pressurized tank contains about 32 times as much air as the non-pressurized tank.

Lesson 7.5

2. $\pm 9, \pm 3, \pm 1, \pm\dfrac{9}{8}, \pm\dfrac{9}{4}, \pm\dfrac{9}{2}, \pm\dfrac{3}{8}, \pm\dfrac{3}{4}, \pm\dfrac{3}{2}, \pm\dfrac{1}{8}, \pm\dfrac{1}{4}, \pm\dfrac{1}{2}$

3.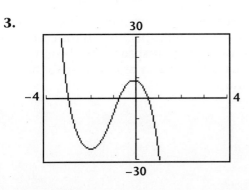

4. roots: $x = -\dfrac{3}{4}$, or $x = \dfrac{1}{2}$, or $x = -3$

5. Since the real roots are less than 1, you can immediately eliminate all possibilities that are greater than 1. Since all the real roots are greater than -4, you can eliminate all possibilities less than -4.

Cooperative-Learning Activity — Chapter 8

Lesson 8.1

1–2.

	Real diameter (km)	Real distance from Sun (km)
Sun	1,391,000	0.00
Mercury	4880	5.8×10^7
Venus	12,100	1.1×10^8
Earth	12,756	1.5×10^8
Mars	6791	2.3×10^8
Jupiter	143,200	7.8×10^8
Saturn	120,000	1.4×10^9
Uranus	51,800	2.9×10^9
Neptune	49,500	4.5×10^9
Pluto	3000	5.9×10^9

	Model diameter (cm)	Model distance from Sun (m)
Sun	25.04	0.0
Mercury	0.09	10.4
Venus	0.22	19.8
Earth	0.23	27.0
Mars	0.12	41.4
Jupiter	2.58	140.4
Saturn	2.16	252.0
Uranus	0.93	522.0
Neptune	0.89	810.0
Pluto	0.05	1062.0

3. If the basketball were 25 centimeters in diameter, Jupiter would be an object that is an inch in diameter, such as a table tennis ball.

4. Pluto would be over 1 kilometer away from your desk.

5. If the basketball were 25 centimeters in diameter, the Earth could be modeled by a pea. It would be located 27 meters (approximately 80 feet) away from your desk.

Algebra 2

ANSWERS

Lesson 8.2

1–5. A. $y = 2 - \dfrac{4}{x+4}$

Reflection across the *x*-axis
Horizontal translation 4 units to the left
Vertical stretch of 4
Vertical translation 2 units up

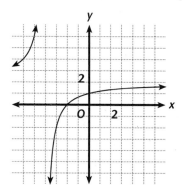

B. $y = -2 + \dfrac{15}{x+3}$

Horizontal translation 3 units to the left
Vertical stretch of 15
Vertical translation 2 units down

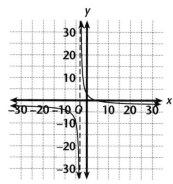

C. $y = 5 + \dfrac{4}{x-1}$

Horizontal translation 1 unit to the right
Vertical stretch of 4
Vertical translation 5 units up

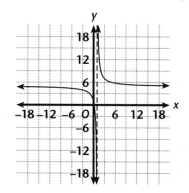

6. Answers may vary. Sample answer: No; the numerator and the denominator must be polynomials of the same degree. If the degrees of the polynomials are different, then the rational function cannot be written as a transformation of $\dfrac{1}{x}$.

Lesson 8.3

2. Answers may vary. Sample answers:
Choose rectangle A: $(x+1)(x-2)$
rectangle B: $(x+1)(x+4)$
Area of rectangle A $= x^2 - x - 2$
Area of rectangle B $= x^2 + 5x + 4$

3. Answers will vary. Sample answer: Using the rectangles in Procedure 2, rectangle B would have the greater area. Choose a comparison factor of 2.
$\dfrac{x^2 + 5x + 4}{x^2 - x - 2} > 2$

4. $x > 2$

5. Check students' work. Answers should be in the same format as 1–4. Students may solve inequalities on a graphics calculator.

Lesson 8.4

1. Known: the speed of the motorboat and the distance from the campsite to the beach; Unknown: the rate of the current and the time required to go up and down the river.

2. $\dfrac{2}{8+c}, \dfrac{2}{8-c}$

3. $\dfrac{2}{8+c} + \dfrac{2}{8-c} = \dfrac{32}{64-c^2}$

4 a. $\dfrac{32}{63}$ hour, $\dfrac{32}{60} = \dfrac{8}{15}$ hour, $\dfrac{32}{55}$ hour

b. The value of *c* cannot be greater than 8 nautical miles per hour. If the current is greater than the speed of the boat, the boat will go backwards relative to the shore.

ANSWERS

Lesson 8.5

2–4.

	r	· t	= d
Father	x	$\frac{1000}{x}$	1000
Mother	$x - 10$	$\frac{500}{x - 10}$	500
Son	$\frac{2x - 10}{2}$	$\frac{500}{x - 5}$	500
Total		33	2000

5. $\frac{1000}{x} + \frac{500}{x - 10} + \frac{500}{x - 5} = 33$

6. $-33x^3 + 2495x^2 - 24{,}150x + 50{,}000$

7. Answers may vary. Sample answer: The zeros can be found on a graphics calculator. A car would probably be traveling at least 40 miles per hour or higher so the maximum value for x should be greater than 40.

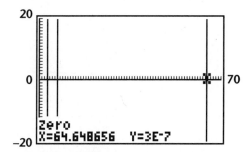

The father's rate is 64.6 miles per hour, the mother's rate is 54.6 miles per hour, and the son's rate is 59.6 miles per hour.

Lesson 8.6

3. Answers may vary. Sample answer: For reflection, vertical translation, $a = -1$, $b = 1$ and $c = -2$: $y = -\sqrt{x - 1} - 2$

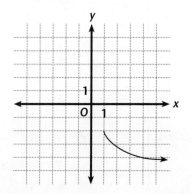

Lesson 8.7

Answers may vary. Sample answer (rows and columns transposed):

Player	1	2
Operation	+	×
A	$\sqrt{4a^4b^2}$	$(27a^3b^3)^{\frac{1}{3}}$
B	$\sqrt{2} - \sqrt{3}$	$\sqrt[3]{8a^3b^3}$
Answer	$2a^2b + \sqrt{2} - \sqrt{3}$	$6a^2b^2$

Lesson 8.8

1. r^3 varies directly as the square of T.

2. $C = 2.5096 \times 10^{19}$;
   $T_p = \sqrt{\frac{(r_p)^3}{C}}$

4. 224.67 Earth days

5. Answers may vary. Sample answer: The average distance between the planet and the Sun.

Cooperative-Learning Activity — Chapter 9

Lesson 9.1

2. Answers may vary. Sample answer: If the city park at $(1, 4)$ and the school at $(-4, -3)$ are chosen, the meeting place is $\left(-\frac{3}{2}, \frac{1}{2}\right)$.

Lesson 9.2

2. The curve is a parabola with point O as the vertex, point F as the focus, and the x-axis as the axis of symmetry.

3. $x = \frac{1}{16}y^2$

ANSWERS

4. Answers may vary. Sample answer: Moving the focus changes the shape of the parabola. As the focus moves further from the vertex, the curve of the parabola becomes less flat.

Lesson 9.3

4. Set B contains nonintersecting circles.

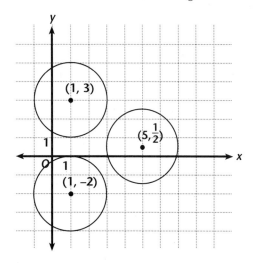

Lesson 9.4

2. Answers may vary. Sample answer:
$$\frac{(x-1)^2}{9} + \frac{(y+2)^2}{16} = 1$$

3. For the equation above: The major axis is parallel to the y-axis with length 8; the minor axis is parallel to the x-axis with length 6.

4. Center: $(1, -2)$
   Foci: $(1, -2 + \sqrt{7}), (1, -2 - \sqrt{7})$
   Vertices: $(1, 2), (1, -6)$
   Covertices: $(-2, -2), (4, -2)$

5.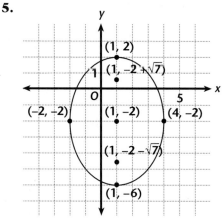

6. Check students' work. At least one ellipse should have its major axis parallel to the y-axis and at least one should have a center other than the origin.

Lesson 9.5

1.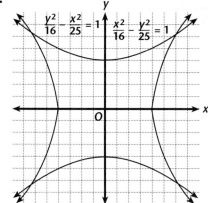

2. Answers may vary. Sample answer: The graphs are the same shape but they open in different directions.

3.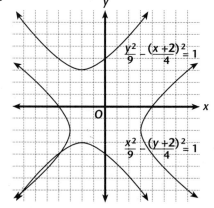

ANSWERS

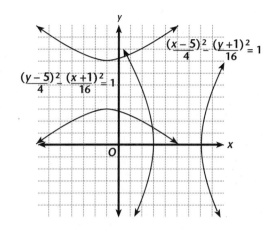

4. Answers may vary. Sample answer: The effect of changing x and y in the equation results in a reflection across the line $y = x$.

Lesson 9.6

2. Distance from Tosha = 1.32 km
   Distance from Emilio = 1.98 km
   Tosha's coordinates: $(-1, 1)$
   Emilio's coordinates: $(-1, -2)$
   $(x + 1)^2 + (y - 1)^2 = (1.32)^2$
   $(x + 1)^2 + (y + 2)^2 = (1.98)^2$
   $(y - 1)^2 - (y + 2)^2 = (1.32)^2 - (1.98)^2$

3. The lightning strike could have had coordinates $(-1.67, -0.14)$ or $(-0.33, -0.14)$.

4.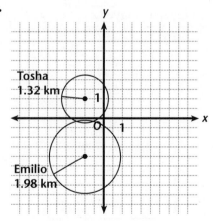

Cooperative-Learning Activity — Chapter 10

Lesson 10.1

1. $\frac{1}{8}$ or 12.5%

0 heads	TTT
1 head	TTH, THT, HTT
2 heads	HHT, HTH, THH
3 heads	HHH

2. Answers may vary. Sample answers:

Outcome	Outcome
TTH	HHT
TTH	HHT
HHT	HTT
HHT	TTT
TTT	HHT
TTT	HHT
HTT	HTH
HHT	THH
HHH	THT
HHH	TTH

3. $\frac{2}{20}$ or 10%;
   Answers may vary. Sample answer: The theoretical probability was higher than the experimental probability. As the number of trials increases, the experimental probability should get closer to the theoretical probability.

Algebra 2 — Answers 111

ANSWERS

Lesson 10.2

1. Answers may vary. Sample answer:
   If the students are A, B, C, and D, the possible arrangements are:

   ABCD ACBD ADBC
   ABDC ACDB ADCB
   BACD BCAD BDAC
   BADC BCDA BDCA
   CABD CBAD CDBA
   CADB CBDA CDAB
   DABC DBCA DCBA
   DACB DBAC DCAB

   Number of arrangements: 24

2. Answers may vary. Sample answer:
   The Photographer keeps one person in the first position until all arrangements of the other three are completed. Then a second person stands in the first position, and the process is repeated.

3. Answers may vary. Sample answer:
   In each existing arrangement of 4 Models, the Photographer can stand to the left of any person or to the right of the last person, so there are five arrangements for each arrangement of 4 Models. Therefore, there are a total of 5 · 24, or 120 different arrangements with the Photographer.

4. 6 different arrangements

Lesson 10.3

1. 20

2. 35

3. 56

4. $\dfrac{n!}{3!(n-3!)}$

5. Answers may vary. Sample answer:
   The number of triangles is the number of combinations of n things taken 3 at a time, because a triangle can be named using its vertices in any order.

Lesson 10.4

1. Answers may vary. Sample tally:

	Male	Female	Total
Rock and roll	1	0	1
Country western	1	0	1
Rap	1	1	2
No favorite	0	0	0
Total	3	1	4

   In 2 and 3, answers will vary.
   Answers are given for the sample tally above.

2. **a.** inclusive; $\frac{3}{4} + \frac{1}{4} - \frac{1}{4} = \frac{3}{4}$

   b. inclusive; $\frac{1}{4} + \frac{1}{4} - \frac{1}{4} = \frac{1}{4}$

3. mutually exclusive; $\frac{1}{4} + \frac{0}{4} = \frac{1}{4}$

Lesson 10.5

2. Answers may vary. Sample answer for goal = 3:
   $P(A) = P(B) = \dfrac{1}{18} \approx 0.06$

   $P(A \text{ and } B) = \dfrac{1}{18} \cdot \dfrac{1}{18} = \dfrac{1}{324} \approx 0.003$

3. Answers may vary. Sample answer:

Goal	Player 1	Player 2
3	10	3
3	2	8
3	4	8
3	7	5
3	2	10
3	9	4
3	11	8
3	11	9
3	5	10
3	4	5

ANSWERS

4. Answers may vary. The experiment agreed with the probability determined by the Probability Expert for P(A and B). In 10 trials, 0 successful outcomes would be expected.

5. Check students' work.

6. Answers may vary. Sample answer: Some goals have a higher probability than others, because there are more possible ways to achieve those goals. The goal with the highest probability is 7 because there are six ways to roll a sum of 7 on two number cubes.

Lesson 10.6

1. Event A is working in agriculture. Event B is working 40 hours per week.

2. $P(A) = \frac{3208}{117,441}$

   $P(A \text{ and } B) = \frac{624}{117,441}$

   $P(B|A) = \frac{624}{3208}$

3. Situation Y: Event A is working in a non-agriculture industry. Event B is working 14 hours or less per week.

   $P(A) = \frac{114,233}{117,441}$

   $P(A \text{ and } B) = \frac{5917}{117,441}$

   $P(B|A) = \frac{5917}{114,233}$

   Situation Z: Event A is working 15 to 29 hours. Event B is working in agriculture.

   $P(A) = \frac{15,115}{117,441}$

   $P(A \text{ and } B) = \frac{493}{117,441}$

   $P(B|A) = \frac{493}{15,115}$

Lesson 10.7

2. Answers may vary. Sample answer:

Trial number	Trial	Number of rolls
1	BCA	3
2	BBAAAC	6
3	BAABC	5
4	BAC	3
5	BAAC	4
6	CACAACAB	8
7	CCBA	4
8	BBBCCA	6
9	CCAB	4
10	BBBAAC	6

3. Check students' work.

4. Answers may vary. Sample answer: For the 10 trials in Procedure 2 above, the mean is 4.9 rolls.

Cooperative-Learning Activity — Chapter 11

Lesson 11.1

2. $a_1 = 18$; $a_2 = 59$

3–4. $a_3 = 182$; $a_4 = 551$; $a_5 = 1658$; $a_6 = 4979$; $a_7 = 14,492$; $a_8 = 44,831$; $a_9 = 134,498$; $a_{10} = 403,499$

5. 604,767

7. $a_1 = -8$; $a_2 = 82$; $a_3 = -278$; $a_4 = 1162$; $a_5 = -4598$; $a_6 = 18,442$; $a_7 = -73,718$; $a_8 = 294,922$; $a_9 = -1,179,638$; $a_{10} = 4,718,602$; $S_{10} = 3,774,970$

ANSWERS

Lesson 11.2

2. a.

Years	0	1	2	3
Value	45,000	41,000	37,000	33,000

4	5	6
29,000	25,000	21,000

b. $t_n = 45,000 - 4000n$, or $t_n = t_{n-1} - 4000$; arithmetic

c.

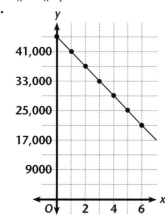

3. a. $t_n = 200n$, or $t_n = t_{n-1} + 200$; arithmetic

b.

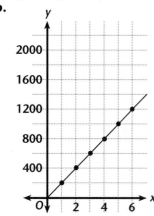

4. Each sequence is defined by adding or subtracting a constant. Thus, each sequence is arithmetic. Each graph contains points that lie on a line.

Lesson 11.3

2. Answers may vary. Sample answer for $n = 8$: $t_8 = -16$, $S_8 = -44$

3. Answers may vary. Sample answer for $n = 14$: $t_{14} = -34$, $S_{14} = -203$

4. Answers may vary. Sample answer for $n = 15$:
Sequence B: $t_{15} = 141$, $S_{15} = 1170$
Sample answer for $n = 10$: $t_{10} = 96$, $S_{10} = 555$
Sequence C: Sample answer for $n = 11$:
$t_{11} = 80$, $S_{11} = 165$
Sample answer for $n = 7$: $t_7 = 28$, $S_7 = -77$

Lesson 11.4

1, 3.

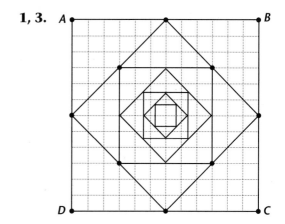

2. length: $6\sqrt{2}$ units; area: 72 square units

3–4. Student measurements may vary.

Exact length	$6\sqrt{2}$	6	$3\sqrt{2}$	3	$1.5\sqrt{2}$	1.5
Area	72	36	18	9	4.5	2.25

5. lengths: geometric, $r = \dfrac{1}{\sqrt{2}}$

area: geometric, $r = \dfrac{1}{2}$

ANSWERS

Lesson 11.5

2. arithmetic: 3, 5
   geometric: 1, 4

3. arithmetic

	t_{10}	S_{10}
sequence 3	−3	60
sequence 5	−136	−640

geometric

	t_{10}	S_{10}
sequence 1	7680	15,345
sequence 4	−4096	−2728

4. arithmetic: 2, 5
   geometric: 3, 4

arithmetic

	t_{10}	S_{10}
sequence 2	−125	−125
sequence 5	9.8	48.5

geometric

	t_{10}	S_{10}
sequence 3	7.508	377.475
sequence 4	-10^8	90,909,090.9

Lesson 11.6

5.

Step number	1	2	3
Step length	$36\left(\frac{1}{2}\right)$	$36\left(\frac{1}{2}\right)^2$	$36\left(\frac{1}{2}\right)^3$

4	5	n
$36\left(\frac{1}{2}\right)^4$	$36\left(\frac{1}{2}\right)^5$	$36\left(\frac{1}{2}\right)^6$

6. Answers may vary. Sample answer: Technically, the Mover will never cross the finish line. The distance can be modeled by $36\left(\frac{1}{2} + \frac{1}{4} + \frac{1}{8} + \cdots\right)$; which would eventually equal 36 if the Mover could continue taking steps forever.

7. The series $36\left(\frac{1}{2} + \frac{1}{4} + \frac{1}{8} + \cdots + \frac{1}{2^n}\right)$ gives the total distance traveled in n steps. The series $36\left(\frac{1}{2} + \frac{1}{4} + \frac{1}{8} + \cdots + \frac{1}{2^n} + \cdots\right)$ gives the total distance the Mover would travel if the Mover could take an infinite number of steps.

$$36\left(\frac{1}{2} + \frac{1}{4} + \frac{1}{8} + \cdots + \frac{1}{2^n} + \cdots\right) =$$

$$36\left(\frac{\frac{1}{2}}{1-\frac{1}{2}}\right) = 36$$

Lesson 11.7

1.
```
                    1   5  10  10   5   1
                1   6  15  20  15   6   1
            1   7  21  35  35  21   7   1
        1   8  28  56  70  56  28   8   1
    1   9  36  84 126 126  84  36   9   1
1  10  45 120 210 252 210 120  45  10   1
```

2. Answers may vary. Sample answer:

Sum: 15

3. Check students' work.

4. Answers may vary. Sample answer: For a diagonal going left to right, the sum can be found in the row after the last number in the ring, to the left of that number. For a diagonal going right to left, the sum can be found in the row after the last number in the ring, to the right of that number.

5.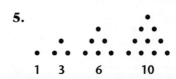

6. 220; The sum corresponds to the number in the next row to the left of the tenth triangular number.

ANSWERS

Lesson 11.8

3. Answers may vary. Sample answer:
   For $\binom{8}{4}$, the polynomial is $(a+b)^8$.
   The specific term is $\frac{8!}{4!4!}a^4b^4$.
   Answers may vary. Sample answers:
   For $\binom{9}{5}$, the polynomial is $(a+b)^9$.
   The specific term is $\frac{9!}{4!5!}a^5b^4$.
   Check students' work for 6 more problems.

6. Check students' work.

Cooperative-Learning Activity — Chapter 12

Lesson 12.1

2. Answers may vary. Sample answer:

Finish bin	Number of finishes
1	2
2	5
3	9
4	9
5	5
6	2

The mean is 3.5, the modes are 3 and 4, and the range is 5.

4.
Finish bin	Number of finishes
1	1
2	3
3	11
4	8
5	8
6	1

The mean is 3.69, the mode is 3, and the range is 5. The mean of the experimental data is close to the prediction, although the frequencies are distributed differently.

Lesson 12.2

1. Answers may vary. Sample answer: The circle graph is a good choice because the data has a meaningful total.

2. Answers may vary. Sample answer: For a circle graph, the percents and degrees of sectors must be calculated.

Year	1950	1997	1950	1997
North America	0.066	0.051	24°	18°
Latin America	0.065	0.085	23°	31°
Europe	0.156	0.087	56°	31°
Asia	0.621	0.644	224°	232°
Africa	0.087	0.128	31°	46°
Oceania	0.005	0.005	2°	2°

3. Answers may vary. Sample answer:

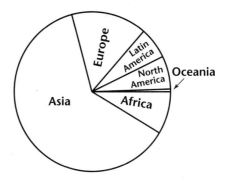

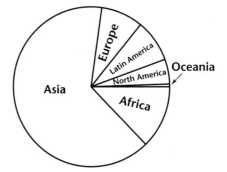

4. Answers may vary. Sample answer: The percent of world population living in North America and Europe decreased, and the percent living in Latin America, Asia, and Africa increased.

ANSWERS

Lesson 12.3

2.
Word length	Number of words
1	3
2	7
3	6
4	17
5	6
6	2
7	0
8	2
9	1

3.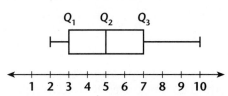

4.
Word length	Number of words
1	0
2	4
3	8
4	7
5	8
6	4
7	4
8	2
9	3
10	3

5. Answers may vary. Sample answer: The first passage is clustered around 3 and 4 letter words. It is more appropriate for younger children. The familiarity of the words and the content should also be considered.

Lesson 12.4

1. 0.746

2.
x_i	$|x_i - 0.75|$
0.750	0.000
0.752	0.002
0.751	0.001
0.692	0.058
0.698	0.052
0.746	0.004
0.749	0.001
0.753	0.003
0.750	0.000
0.745	0.005
Total	0.126

x_i	$|x_i - 0.75|$
0.748	0.002
0.751	0.001
0.753	0.003
0.755	0.005
0.748	0.002
0.746	0.004
0.749	0.001
0.754	0.004
0.750	0.000
0.751	0.001
Total	0.023

Algebra 2   Answers **117**

ANSWERS

x_i	$\|x_i - 0.75\|$
0.752	0.002
0.753	0.003
0.750	0.000
0.748	0.002
0.747	0.003
0.749	0.001
0.750	0.000
0.752	0.002
0.753	0.003
0.748	0.002
Total	0.018

3. The mean deviations for the three data sets above are 0.0126, 0.0023, and 0.0018. The mean deviation for the whole set is 0.0056, which is greater than 0.001. The batch should not be released.

Lesson 12.5

1. Answers may vary. Sample answer: This is a binomial experiment because the trials are independent and there are exactly two outcomes (being chosen and not being chosen) whose probabilities remain constant. Each day represents a trial, so $n = 10$. The successful outcomes are exactly 3, 4, 5, 6, 7, 8, 9, and 10 days.

2. $p = \frac{1}{7} \approx 0.14; q \approx 1 - 0.14 = 0.86$

3. $_7C_3(0.14)^3(0.86)^4$

5. $P(\text{exactly } 3) = 0.0525$

6. Answers may vary. Sample answer: The most efficient way to find $P(\text{at least } 3)$ is to find $P(\text{exactly } 0)$, $P(\text{exactly } 1)$, $P(\text{exactly } 2)$, add these probabilities, and subtract the sum from 1.
$P(\text{exactly } 0) = {_7C_0}(0.14)^0(0.86)^7 \approx 0.3479$
$P(\text{exactly } 1) = {_7C_1}(0.14)^1(0.86)^6 \approx 0.3965$
$P(\text{exactly } 2) = {_7C_2}(0.14)^2(0.86)^5 \approx 0.1936$
$P(\text{at least } 3) = 1 - (0.3479 + 0.3965 + 0.1936) = 0.062$

Lesson 12.6

2.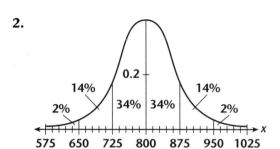

3. **a.** 875
   b. 800
   c. 725
   d. 650

4. The Store Manager should order 950 cases.

5. Answers may vary. Sample answer: The advertising special may mean that the sales no longer fall in a normal distribution. The store manager should order extra stock—even more than 950 cases.

Cooperative Learning Activity — Chapter 13

Lesson 13.1

2.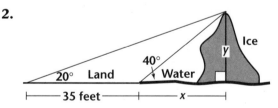

4. $\tan 40° = \dfrac{y}{x}$

   $\tan 20° = \dfrac{y}{35 + x}$

5. about 22.5 feet

6. Answers may vary. Sample answer: This method could be used to determine the height of a mountain. Two different angles of elevation could be used from two points that are located a known distance apart.

ANSWERS

Lesson 13.2

Answers may vary. Sample answer for II and $|\cos \theta| = \frac{\sqrt{3}}{2}$:

$\cos \theta = -\frac{\sqrt{3}}{2}$, $\sin \theta = \frac{1}{2}$, $\tan \theta = -\frac{\sqrt{3}}{3}$

$\sec \theta = -\frac{2\sqrt{3}}{3}$, $\csc \theta = 2$, $\cot \theta = -\sqrt{3}$

Lesson 13.3

3. Answers may vary. Sample answer: The continuation of the graph of
$$y = \begin{cases} 1 & \text{if } 0 \leq x < 1 \\ -x + 2 & \text{if } 1 \leq x < 3 \\ -1 & \text{if } 3 \leq x < 4 \end{cases}$$ given that it is periodic with a period of 4 is shown below.

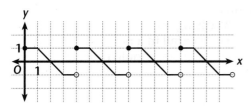

5. Answers may vary. Sample answer: If the entire graph can be made by repeating one part of it, you can conclude that the graph represents a periodic function.

Lesson 13.4

1.

	Central angle	Radius of circle	Radian measure	Arc length
a.	180°	2 inches	π	2π
b.	90°	2 inches	$\frac{\pi}{2}$	π
c.	45°	2 inches	$\frac{\pi}{4}$	$\frac{\pi}{2}$
d.	30°	2 inches	$\frac{\pi}{6}$	$\frac{2\pi}{3}$

2.

	Central angle	Radius of circle	Radian measure	Arc length
a.	180°	4 inches	π	4π
b.	90°	4 inches	$\frac{\pi}{2}$	2π
c.	45°	4 inches	$\frac{\pi}{4}$	π
d.	30°	4 inches	$\frac{\pi}{6}$	$\frac{4\pi}{3}$

3. Answers may vary. Sample answer: The arc length is proportional to the radius and to the radian measure of the angle.

4.

	Central angle	Radius of circle	Degree measure	Arc length
a.	$\frac{3\pi}{2}$	3 inches	270°	$\frac{9\pi}{2}$
b.	$\frac{7\pi}{4}$	2 inches	315°	$\frac{7\pi}{2}$
c.	$\frac{7\pi}{6}$	5 inches	210°	$\frac{35\pi}{6}$

5. Answers may vary. Sample answer: Degree measure could be used if you set up a proportion using $\frac{\text{degree measure}}{360}$. It is much easier to use radians and the formula $S = r\theta$.

Lesson 13.5

2. Function A has only an amplitude change from the parent function.

3. amplitude 3; period 2π; 0 phase shift

4.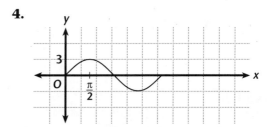

5.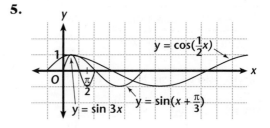

ANSWERS

Lesson 13.6

Answers may vary. Sample answer:

3.

Object height	Length of shadow	Angle of depression
6 inches	15 inches	21.8°
6 inches	6 inches	45°
6 inches	4 inches	56.3°
6 inches	3 inches	63.4°
6 inches	2 inches	71.6°

5. Answers may vary. Sample answer: As the length of the shadow decreases, the angle of depression increases.

6. Answers may vary. Sample answer: The flashlight models the movement of the Sun as the day progresses. The shadows decrease until noon, when the Sun is directly overhead and the angle of elevation is 90°. As evening ensues, the Sun lowers, the angle of depression decreases, and the shadows lengthen.

Cooperative Learning — Chapter 14

Lesson 14.1

3. Answers may vary. Sample answer:

AB	m∠A	m∠B
40 inches	50°	70°

4. $\dfrac{\sin A}{a} = \dfrac{\sin C}{c}$, $\dfrac{\sin B}{b} = \dfrac{\sin C}{c}$

$\dfrac{\sin 50°}{a} = \dfrac{\sin 60°}{40}$, $\dfrac{\sin 70°}{b} = \dfrac{\sin 60°}{40}$

5. $a \approx 35.4$ inches $b \approx 43.4$ inches

6. $\dfrac{\sin 52°}{a} = \dfrac{\sin 61°}{42}$, $\dfrac{\sin 67°}{b} = \dfrac{\sin 61°}{42}$
$a \approx 35.4$ inches $b \approx 43.4$ inches

Lesson 14.2

2. SAS

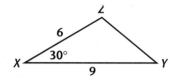

3. $x^2 = 6^2 + 9^2 - 2 \cdot 6 \cdot 9 \cos 30° \approx 23.5$
$a \approx 4.8$
$\dfrac{\sin 30°}{4.8} = \dfrac{\sin Z}{9}$, m∠$Z = 110.4°$
m∠$Y = 180° - 30° - 110.4° = 39.6°$

You could also apply the law of cosines to find m∠Y.

4. ASA; m∠$X = 60°$
$\dfrac{\sin 60°}{9} = \dfrac{\sin 70°}{z}$; $z \approx 9.8$
$\dfrac{\sin 60°}{9} = \dfrac{\sin 50°}{y}$; $y \approx 8.0$

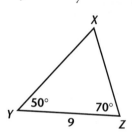

5. C. SSA
$\dfrac{\sin 20°}{2} = \dfrac{\sin X}{4}$, m∠$X \approx 43.2°$ or $136.8°$
so m∠$Z = 116.8°$ or $23.2°$
The law of cosines must be used to find both values of z.
$z^2 = x^2 + y^2 - 2xy \cos Z$
$z^2 = 4^2 + 2^2 - 2 \cdot 4 \cdot 2 \cos 23.2°$, $z \approx 2.3$
or $z^2 = 4^2 + 2^2 - 2 \cdot 4 \cdot 2 \cos 116.8°$,
$z \approx 5.2$

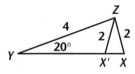

D. not possible

6. Depending on the information that is given or found, you may have many choices of laws and facts that can be used to find more information.

ANSWERS

Lesson 14.3

Answers may vary. Sample answer:

2. A.

| $\tan\theta$ | $\dfrac{\sin\theta}{\cos\theta}$ | $\dfrac{1}{\cot\theta}$ |

3. B.

| $\dfrac{\sec\theta}{\csc\theta}$ | $\dfrac{\sin\theta}{\cos\theta}$ | $\tan\theta$ |

4.

C. $1 - \sin^2\theta$	XXX	$\cos^2\theta$
D. $\cot\theta\sec\theta$	$\dfrac{1}{\sin\theta}$	$\csc\theta$
E. $\tan^2\theta + 1$	$\dfrac{1}{\cos^2\theta}$	$\sec^2\theta$
F. $\dfrac{\tan\theta}{\cot\theta}$	$\dfrac{\sin^2\theta}{\cos^2\theta}$	$\dfrac{1-\cos^2\theta}{\cos^2\theta}$

5. Answers may vary. Sample answer: By writing the expressions in terms of sine and cosine, it may be possible to use more identities to simplify. By writing the expressions in terms of one function, it may be possible to combine terms and simplify further.

Lesson 14.4

1.

$n°$	$\sin n°$	$\cos n°$
$n° = 2°$	0.0350	0.9993
$n° = 3°$	0.0525	0.9985
$n° = 4°$	0.0670	0.9975
$n° = 5°$	0.0844	0.9901

2. To fill in one row of the table for the sine, use sin 1° and cos 1° along with the row of the table just filled. Use the values in the addition formula for the sine. Similarly, use the addition formula for the cosine to fill in the cosine part of the row.
$\sin(n+1)° = 0.9998 \sin n° + 0.9998 \cos n°$;
$\cos(n+1)° = 0.9998 \cos n° - 0.0175 \sin n°$;
Each of these equations are recursive formulas for sequences.

3. Because the trigonometric function values used are rounded, one can expect the rounding error to become more significant as n increases.

4. The additional information needed is $\sin\left(k+\dfrac{1}{2}\right)°$ and $\cos\left(k+\dfrac{1}{2}\right)°$ for some integer k, such as 0.

Lesson 14.5

2 a. $\sin\left(\dfrac{\theta}{2}\right) = \dfrac{\frac{s}{2}}{a}$

b. $\sqrt{\dfrac{1-\cos\theta}{2}} = \dfrac{\frac{s}{2}}{a}$

c. Use + because $\dfrac{\theta}{2}$ is between 0° and 90°.

3. Answers may vary. Sample answer: $\theta = 90°$ and $s = 48$ inches.

4. $a = 24\sqrt{2}$, about 33.9 inches

5. Answers may vary. Sample answer for $\theta = 60°$ and $s = 40$ inches: $a = 40$ inches

6. $\theta = \cos^{-1}\left(\dfrac{2a^2 - s^2}{2a^2}\right)$

7. The equation in Procedure 6 can be used to determine the roof angle when roof length and doorway width are given.

Lesson 14.6

2.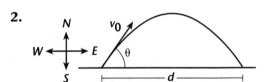

3. 22.8° or 67.2°

4. Answers may vary. Sample answer: The angle would have to be the complement of the original angle.

5. 17.4° or 72.6°

6. The maximum distance is achieved when $\sin 2\theta = 1$, that is, when $\theta = 45°$.

Algebra 1
Geometry
Algebra 2

HOLT, RINEHART AND WINSTON
A Harcourt Classroom Education Company